Group Identities on Units and Symmetric Units of Group Rings

Algebra and Applications

Volume 12

Algebra and Applications aims to publish well written and carefully refereed monographs with up-to-date information about progress in all fields of algebra, its classical impact on commutative and noncommutative algebraic and differential geometry, K-theory and algebraic topology, as well as applications in related domains, such as number theory, homotopy and (co)homology theory, physics and discrete mathematics.

Particular emphasis will be put on state-of-the-art topics such as rings of differential operators, Lie algebras and super-algebras, group rings and algebras, C^*-algebras, Kac-Moody theory, arithmetic algebraic geometry, Hopf algebras and quantum groups, as well as their applications. In addition, Algebra and Applications will also publish monographs dedicated to computational aspects of these topics as well as algebraic and geometric methods in computer science.

Gregory T. Lee

Group Identities on Units and Symmetric Units of Group Rings

Springer

Gregory T. Lee
Department of Mathematical Sciences
Lakehead University
Thunder Bay, Ontario P7B 5E1
Canada

ISBN 978-1-4471-2589-1   ISBN 978-1-84996-504-0 (eBook)
DOI 10.1007/978-1-84996-504-0
Springer London Dordrecht Heidelberg New York

British Library Cataloguing in Publication Data
A catalogue record for this book is available from the British Library

AMS Codes: 16S34, 16U60, 20C07, 16W10, 16R50

Cover design: SPi Publisher Services

Printed on acid-free paper

Springer is part of Springer Science+Business Media (www.springer.com)

For Michelle

Preface

The unit group $\mathcal{U}(FG)$ of the group ring FG is an important object of study. It is interesting to explore the conditions under which this group satisfies various identities. Early results of this type appear in Sehgal's classic book [94]. My purpose here is to pick up where that book leaves off and present the fascinating results that have appeared, for the most part, since the mid nineties.

A group G (or a subset S of G) is said to satisfy a group identity if there is a nontrivial reduced word $w(x_1,\dots,x_n)$ in the free group $\langle x_1,x_2,\dots \rangle$ such that $w(g_1,\dots,g_n) = 1$ for all $g_i \in G$ (or in S, as the case may be). A series of papers by Giambruno et al. [31, 38, 40], Liu [75], Liu and Passman [76] and Passman [84] presented the conditions under which $\mathcal{U}(FG)$ satisfies a group identity.

But we can also consider the involution $*$ on FG given by

$$\left(\sum_{g \in G} \alpha_g g \right)^* = \sum_{g \in G} \alpha_g g^{-1}.$$

Denote by $\mathcal{U}^+(FG)$ the set of symmetric units; that is, $\mathcal{U}^+(FG) = \{\alpha \in \mathcal{U}(FG) : \alpha^* = \alpha\}$. There have been quite a few papers since the late nineties devoted to discovering the extent to which the symmetric units determine the group structure of $\mathcal{U}(FG)$. In particular, Giambruno et al. [39] and Sehgal and Valenti [96] determined when $\mathcal{U}^+(FG)$ satisfies a group identity. Furthermore, a series of papers by Lee [67], Lee et al. [69] and Lee and Spinelli [73, 74] examined the conditions under which $\mathcal{U}^+(FG)$ satisfies specific group identities. The quaternion group Q_8 tends to provide a counterexample to every question asked, since it is easily verified that the symmetric elements of FQ_8 are central. It is, however, an engrossing problem to classify the counterexamples.

In an effort to maintain consistency, I always assume that our group ring is over a field, even though some of the results have been extended to suitable domains or other rings. Also, the involution on FG will always be the classical one described above, although I break this rule in Chapter 7 in order to mention a few recent results allowing more general involutions.

I assume that the reader has had an introductory graduate course on groups and rings, and so use standard results freely. Some prior exposure to the concept of a group ring (in particular, to the famous books of Passman [82] and Sehgal [94]) would be an asset, but it is not essential. To be sure, I am not attempting anything so ambitious as these books, where essentially, all of the major results concerning the ring structure of FG and the group structure of $\mathscr{U}(FG)$ were discussed. Given the explosion of work on group rings in recent years, such an attempt now would be all but impossible. (It would most definitely be impossible for this author!) Instead, this book is largely a continuation of the chapters *Lie Properties in KG* and *Units in Group Rings II* in [94]. I freely borrow theorems from [82] and [94], but the purloined results are all stated in full for the reader's convenience. Also, it will be clear to the reader that results labeled "Theorem" in this book are the main theorems on group identities and Lie identities of group rings. A number of major classical results are, therefore, presented as propositions.

This book contains seven chapters and an appendix. Here is a brief summary of their contents.

In Chapter 1, we discuss the conditions under which $\mathscr{U}(FG)$ satisfies a group identity. The first step is to establish Hartley's conjecture; namely, if $\mathscr{U}(FG)$ satisfies a group identity, and G is torsion, then FG satisfies a polynomial identity. After that, necessary and sufficient conditions are found, for both torsion and nontorsion groups G. When G is nontorsion, a suitable restriction must be imposed upon G modulo its torsion part for the sufficiency. A well-known conjecture due to Kaplansky states that if G is torsion-free, then the only units of FG are the trivial units, namely λg, with $0 \neq \lambda \in F$ and $g \in G$. It is certainly true if G is a u.p. (unique product) group, and we impose this as our restriction upon G modulo its torsion part.

Chapter 2 covers the same territory for the symmetric units. We find that if F is infinite and G is torsion, then FG satisfies a polynomial identity whenever $\mathscr{U}^+(FG)$ satisfies a group identity. Necessary and sufficient conditions are then presented for $\mathscr{U}^+(FG)$ to satisfy a group identity, for both torsion and nontorsion groups G.

Chapter 3 contains proofs of theorems concerning Lie identities satisfied by the set of symmetric elements, $(FG)^+$. (The corresponding results for all of FG are classical.) These results are essential to our later discussion, as there is an intimate connection between the Lie properties of FG (resp., $(FG)^+$), and corresponding group identities of $\mathscr{U}(FG)$ (resp., $\mathscr{U}^+(FG)$).

In Chapters 4, 5 and 6, we consider particular group identities. Chapter 4 contains classical results establishing when $\mathscr{U}(FG)$ is nilpotent and recent results for $\mathscr{U}^+(FG)$. Similarly, Chapters 5 and 6 contain results pertaining to the bounded Engel and solvability properties respectively, for both $\mathscr{U}(FG)$ and $\mathscr{U}^+(FG)$.

In Chapter 7, some related results are mentioned. In particular, we discuss some identities related to those mentioned earlier, and some recent results concerning involutions other than the classical one on FG.

The appendix contains some results concerning the central closure of a prime ring. These are needed in order to prove Proposition 2.2.2.

I would like to thank Ernesto Spinelli and the anonymous reviewers for their helpful suggestions. I also thank Lynn Brandon and Lauren Stoney at Springer for their help with the publication process. Portions of this book were written while I was visiting the University of Alberta and the Università del Salento, and I thank their respective mathematics departments for their warm hospitality. Also, I want to thank my wife and family for their ongoing support. Finally, I thank my teacher, Professor Sudarshan Sehgal, who introduced me to this beautiful subject years ago, and who has been a tremendous source of help to me ever since.

Thunder Bay, Ontario
February 2010   Gregory T. Lee

Contents

Chapter 1
Group Identities on Units of Group Rings

1.1 Introduction

Let G be a group and R a ring. (Unless otherwise indicated, we shall assume throughout the book that every ring has an identity.) The group ring RG is the set of all formal sums

$$\sum_{g \in G} r_g g$$

with each $r_g \in R$ and all but finitely many $r_g = 0$. It forms a ring under the operations

$$\left(\sum_{g \in G} r_g g \right) + \left(\sum_{g \in G} s_g g \right) = \sum_{g \in G} (r_g + s_g) g$$

and

$$\left(\sum_{g \in G} r_g g \right) \left(\sum_{g \in G} s_g g \right) = \sum_{g \in G} \left(\sum_{h \in G} r_h s_{h^{-1} g} \right) g.$$

We ignore the zero terms in an element and just write $r_1 g_1 + \cdots + r_n g_n$. In particular, we identify the elements of G with the group ring elements $1g$ and the elements of R with the group ring elements $r1$.

If N is a normal subgroup of G, then the natural homomorphism $G \to G/N$ induces a ring homomorphism $\varepsilon_N : RG \to R(G/N)$. We write $\Delta(G,N)$ for the kernel of this homomorphism. It is easily seen to be the ideal consisting of the finite sums of terms of the form $rg(n-1)$, with $r \in R$, $g \in G$, and $n \in N$. In particular, if $N = G$, then $\varepsilon = \varepsilon_G$ is called the augmentation map, and we write $\Delta(G) = \Delta(G,G)$. The ideal $\Delta(G)$ is called the augmentation ideal. This is the set of sums of terms of the form $r(g-1)$, with $r \in R$ and $g \in G$. Let us mention a well-known result.

Lemma 1.1.1. *Let G be a group and R a commutative ring of characteristic p^m, for some prime p. If N is a finite normal subgroup of G, then $\Delta(G,N)$ is a nilpotent ideal if and only if N is a p-group.*

G.T. Lee, *Group Identities on Units and Symmetric Units of Group Rings*,
Algebra and Applications 12, DOI 10.1007/978-1-84996-504-0_1,
© Springer-Verlag London Limited 2010

Proof. Suppose N is a finite p-group. If $\alpha_1, \ldots, \alpha_l \in \Delta(G,N)$, then we can see that $\alpha_1 \cdots \alpha_l$ is a sum of terms of the form

$$rg_1(n_1 - 1)g_2(n_2 - 1) \cdots g_l(n_l - 1),$$

with each $g_i \in G$, $n_i \in N$ and $r \in R$. As N is normal, this can be rewritten as

$$rg(n_1' - 1)(n_2' - 1) \cdots (n_l' - 1),$$

with $g \in G$ and each $n_i' \in N$.

That is, it suffices to show that $\Delta(N)$ is nilpotent. Our proof is by induction on the order of N. The result is trivial if $|N| = 1$. Otherwise, we know that N has an element z of order p in its centre. Then by induction, there exists a k such that $(\Delta(N/\langle z \rangle))^k = 0$. That is, $(\Delta(N))^k \subseteq \Delta(N, \langle z \rangle)$. But it is easy to see that $\Delta(N, \langle z \rangle) = (z-1)RN$, and as z is central in RN and has order p, we have $(z-1)^p \in pR\langle z \rangle$; hence $(\Delta(N))^{kpm} \subseteq (z-1)^{pm}RN = 0$, as required.

Conversely, if N is not a p-group, then let $h \in N$ be a p'-element. Then $h - 1 \in \Delta(G,N)$; but if $(h-1)^n = 0$, then $(h-1)^{p^j} = 0$ for some suitably large j. However, $(h-1)^{p^j} \equiv h^{p^j} - 1 \pmod{pRN}$, and $h^{p^j} - 1 \notin pRN$, giving us a contradiction. $\square$

We will largely be interested in the group ring FG over a field F, and in particular the unit group, $\mathcal{U}(FG)$. This is the set of all elements $u \in FG$ for which there exists $v \in FG$ such that $uv = vu = 1$. Aside from the trivial units λg, with $0 \neq \lambda \in F$ and $g \in G$, constructing units can be a difficult problem. Of course, in any ring R, if $r \in R$ satisfies $r^n = 0$, for some positive integer n, then

$$(1+r)(1 - r + r^2 - \cdots \pm r^{n-1}) = 1.$$

Thus, $1 + r \in \mathcal{U}(R)$.

We can apply this to group rings in the following way. If H is a finite subgroup of G, let

$$\hat{H} = \sum_{h \in H} h.$$

In particular, if g is a torsion element of G, let

$$\hat{g} = \widehat{\langle g \rangle} = \sum_{i=0}^{o(g)-1} g^i.$$

Now, suppose that N is a finite normal subgroup of G, with $|N|$ divisible by char F. Then clearly $(\hat{N})^2 = |N|\hat{N} = 0$, and $\hat{N}$ is central. Thus, for any $\alpha \in FG$,

$$(1 + \alpha\hat{N})(1 - \alpha\hat{N}) = 1.$$

Let $\langle x_1, x_2, \ldots \rangle$ be the free group on a countable infinitude of generators. If G is any group, then we say that G satisfies a group identity if there exists a nontrivial reduced word $w(x_1, \ldots, x_n) \in \langle x_1, x_2, \ldots \rangle$ such that $w(g_1, \ldots, g_n) = 1$ for all $g_i \in G$.

To give some examples of group identities, define

$$(x_1, x_2) = x_1^{-1} x_2^{-1} x_1 x_2$$

and recursively,

$$(x_1, \ldots, x_{n+1}) = ((x_1, \ldots, x_n), x_{n+1}).$$

Then G is abelian if it satisfies

$$(x_1, x_2) = 1$$

or nilpotent if it satisfies

$$(x_1, x_2, \ldots, x_n) = 1$$

for some $n \geq 2$. We also say that it is n-Engel if it satisfies

$$(x_1, \underbrace{x_2, x_2, \ldots, x_2}_{n \text{ times}}) = 1$$

and bounded Engel if it is n-Engel for some n.

The following reduction will be useful.

Lemma 1.1.2. *If a group G satisfies a group identity, then it also satisfies a group identity of the form*

$$x^{a_1} y^{b_1} x^{a_2} y^{b_2} \cdots x^{a_n} y^{b_n} = 1,$$

where each a_i and b_i is nonzero, $a_1 < 0$ and $b_n > 0$.

Proof. Suppose that G satisfies $w(x_1, \ldots, x_n) = 1$. Substitute $x^i y x^{-i}$ for each x_i, and notice that $(x^i y x^{-i})^j = x^i y^j x^{-i}$ for any integer j. Reducing, we obtain a nontrivial group identity

$$x^{c_1} y^{d_1} \cdots x^{c_k}$$

for G, with each c_i and d_i nonzero. Thus, conjugating by x^{-c_k}, we see that G also satisfies

$$x^{c_1 + c_k} y^{d_1} \cdots y^{d_{k-1}}.$$

If $c_1 + c_k \neq 0$, then we are essentially done (replacing x with x^{-1} and y with y^{-1} if necessary, in order to make the first exponent negative and the last positive). Otherwise, we have the group identity

$$y^{d_1} x^{c_2} \cdots y^{d_{k-1}}.$$

Interchange x and y and repeat. Eventually we either obtain an identity of the correct form, or else we end up with the identity $x^m = 1$ for some positive integer m. But in the latter case, replacing x with $x^{-1} y$ does the job. □

The earliest results about units in group rings satisfying a group identity concerned specific identities, such as nilpotency and solvability; we will discuss these in later chapters. We begin with the most general situation. Brian Hartley made the following famous conjecture.

Conjecture 1.1.3. Let G be a torsion group and F a field. If $\mathscr{U}(FG)$ satisfies a group identity, then FG satisfies a polynomial identity.

By this we mean that if we let $F\{x_1, x_2, \ldots\}$ denote the free algebra on noncommuting indeterminates $x_1, x_2, \ldots$, then there exists a nonzero polynomial $f(x_1, \ldots, x_n)$ in $F\{x_1, x_2, \ldots\}$ such that $f(\alpha_1, \ldots, \alpha_n) = 0$ for all $\alpha_i \in FG$. For example, a commutative ring would satisfy $x_1 x_2 - x_2 x_1 = 0$. The conditions under which FG satisfies a polynomial identity were determined in classical results due to Isaacs and Passman. Recall that for any prime p, a group G is said to be p-abelian if its commutator subgroup G' is a finite p-group and that 0-abelian means abelian.

Proposition 1.1.4. *Let F be a field of characteristic $p \geq 0$ and G a group. Then FG satisfies a polynomial identity if and only if G has a p-abelian subgroup of finite index.*

Proof. See [82, Corollaries 5.3.8 and 5.3.10]. □

Assume first that the field is infinite. Gonçalves and Mandel [43] verified Hartley's conjecture in the special case that the group identity is actually a semigroup identity (that is, an identity of the form $x_{i_1} x_{i_2} \cdots x_{i_k} = x_{j_1} x_{j_2} \cdots x_{j_l}$). Giambruno et al. handled the characteristic 0 case as well as the characteristic $p > 0$ case where G has no p-elements in [31]. The general case was proved by Giambruno et al. in [38]. Passman then found necessary and sufficient conditions for $\mathscr{U}(FG)$ to satisfy a group identity in [84].

When F is finite, the result is different. Liu verified Hartley's conjecture for finite fields in [75], and Liu and Passman found necessary and sufficient conditions in [76].

In the next section, we present the proof of the conjecture, and the proof of the necessary and sufficient conditions is given in the following section. The remainder of the chapter is devoted to the result of Giambruno et al. [40] classifying the groups with elements of infinite order such that $\mathscr{U}(FG)$ satisfies a group identity.

1.2 Proof of Hartley's Conjecture

Throughout, we let F be a field of characteristic $p \geq 0$. Let us discuss a few facts about group identities on $\mathscr{U}(R)$, where R is an F-algebra. It will be useful, in particular, to consider the case where $R = M_n(F)$, the ring of $n \times n$ matrices over F. For instance, if G is a finite group, and p does not divide $|G|$, then by Maschke's theorem (see [62, Theorem 6.1]), FG is semisimple. It is, of course, Artinian as well, since it is finite-dimensional. Thus, we can apply the Wedderburn–Artin theorem, namely

Proposition 1.2.1. *Let R be a semisimple Artinian ring. Then*

$$R \cong M_{n_1}(D_1) \oplus \cdots \oplus M_{n_k}(D_k),$$

where each n_i is a positive integer and each D_i is a division ring.

Proof. See [62, Theorem 3.5]. □

But we will be able to concern ourselves only with the case where the division algebra is a field, due to the following result, which is Lemma 2.0 in Gonçalves [42].

Proposition 1.2.2. *Let D be a noncommutative division algebra such that D is finite-dimensional over its centre. Then $\mathscr{U}(D)$ contains a nonabelian free group.*

As a nonabelian free group cannot possibly satisfy a group identity, we can dispense with this case.

Let us fix a group identity of the form $w(x,y) = 1$ described in Lemma 1.1.2. We begin with

Lemma 1.2.3. *There exists a nonzero polynomial $f(x)$ such that for every F-algebra R whose unit group satisfies $w(x,y) = 1$, we have $f(ab) = 0$ for every a and b in R such that $a^2 = b^2 = 0$. This polynomial has coefficients in the prime subfield of F and depends only upon w and the characteristic of F.*

Proof. Substituting $x = x_2 x_1^{-1}$ and $y = x_1^{-1} x_2$ in w, we obtain a new group identity for $\mathscr{U}(R)$, of the form

$$1 = v(x_1, x_2) = x_1^{m_1} x_2^{n_1} \cdots x_1^{m_k} x_2^{n_k},$$

with each $m_i, n_i \in \{\pm 1, \pm 2\}$ and $m_1 = n_k = 1$.

Suppose that char $F \neq 2$. Define a polynomial

$$g(x_1, x_2) = (1 + m_1 x_1)(1 + n_1 x_2) \cdots (1 + m_k x_1)(1 + n_k x_2) - 1.$$

Notice that $g(x_1, x_2) = g_1(x_1, x_2) + g_2(x_1, x_2) + g_3(x_1, x_2)$, where g_1 is the sum of all of the monomials in g in which either x_1^2 or x_2^2 appears, g_2 is the sum of the other monomials starting with x_1 and ending with x_2, and g_3 is the sum of the remaining terms. Now, g_2 is a linear combination of terms of the form $(x_1 x_2)^i$ for various i (otherwise, an x_1^2 or x_2^2 must appear), and, indeed, the unique term of highest degree in g is $m_1 n_1 \cdots m_k n_k (x_1 x_2)^k$, a monomial in g_2. As the characteristic is not 2, this is not a zero term.

Now, $x_2 g_2(x_1, x_2) x_1$ is a polynomial in $x_2 x_1$, so let us take $f(x)$ such that $f(x_2 x_1) = x_2 g_2(x_1, x_2) x_1$. Notice that $g_1(a, b) = 0$, since a^2 or b^2 will appear, and both are zero. Also, $b g_3(a, b) a = 0$, since each monomial in g_3 starts with x_2 or ends with x_1, hence we will have b^2 at the beginning or a^2 at the end.

Thus, $f(ba) = b g_2(a, b) a = b g(a, b) a$, so it suffices to show that $b g(a, b) a = 0$. But notice that for any integer n, $(1 + a)^n = 1 + na$, and similarly for b, so in particular, $1 + a$ and $1 + b$ are units, and

$$\begin{aligned}
1 &= v(1 + a, 1 + b) \\
&= (1 + a)^{m_1}(1 + b)^{n_1} \cdots (1 + a)^{m_k}(1 + b)^{n_k} \\
&= (1 + m_1 a)(1 + n_1 b) \cdots (1 + m_k a)(1 + n_k b) \\
&= g(a, b) + 1,
\end{aligned}$$

hence $g(a,b) = 0$, as required.

Suppose now that char $F = 2$. Replacing x_1 with x_1x_2 and x_2 with x_1x_3 in $v(x_1,x_2)$ and reducing, we may instead assume that our group identity is of the form

$$v(x_1,x_2,x_3) = x_{i_1}^{l_1} \cdots x_{i_r}^{l_r},$$

with $1 \leq i_j \leq 3$ for all j, $i_j \neq i_{j+1}$, and each $l_j \in \{\pm 1\}$. Also, $i_1 = 1$ and $i_r = 3$. Define a polynomial

$$g(x_1,x_2) = (1+y_1)\cdots(1+y_r) - 1,$$

where

$$y_j = \begin{cases} x_1x_2x_1, & \text{if } i_j = 1, \\ x_2x_1x_2x_1x_2, & \text{if } i_j = 2, \\ (x_1+x_2x_1)x_2(x_1+x_1x_2), & \text{if } i_j = 3. \end{cases}$$

Now write $g = g_1 + g_2 + g_3$ as in the preceding case. Calculating $y_1 \cdots y_r$, and noting that $y_1 = x_1x_2x_1$ and $y_r = (x_1+x_2x_1)x_2(x_1+x_1x_2)$, we see that g_2 has a unique term $(x_1x_2)^s$ of highest degree. In particular, g_2 is not the zero polynomial and again, $x_2g_2(x_1,x_2)x_1$ is a polynomial in x_2x_1, say $f(x)$. Furthermore, $aba, babab$ and $(a+ba)b(a+ab)$ are all square-zero. (The first two of these are obvious. For the third, notice that

$$(a+ab)(a+ba) = a^2 + 2aba + ab^2a = 0,$$

since $a^2 = b^2 = 0$ and char $F = 2$.) Thus,

$$(1+aba)^2 = (1+babab)^2 = (1+(a+ba)b(a+ab))^2 = 1.$$

Therefore,
$$v(1+aba, 1+babab, 1+(a+ba)b(a+ab)) = 1.$$

It follows that

$$0 = b(v(1+aba, 1+babab, 1+(a+ba)b(a+ab)) - 1)a = bg(a,b)a$$

and, as in the previous case, we see that this implies that

$$0 = bg_2(a,b)a = f(ba).$$

We are done.   □

It will be helpful later to express the next lemma in slightly greater generality than is presently necessary; that is, in terms of domains, rather than fields.

Lemma 1.2.4. *Let D be an integral domain, and let $f(x) \in D[x]$ be a polynomial of degree $n \geq 1$. Let R be a D-algebra, and suppose that D has no zero divisors in R. If $r \in R$, and there exist distinct $\lambda_i \in D$, $1 \leq i \leq n+1$, such that $f(\lambda_i r) = 0$ for all i, then $r^n = 0$.*

Proof. Let $f(x) = a_0 + a_1 x + \cdots + a_n x^n$. Then we get

$$\begin{pmatrix} 1 & \lambda_1 & \lambda_1^2 & \cdots & \lambda_1^n \\ 1 & \lambda_2 & \lambda_2^2 & \cdots & \lambda_2^n \\ \vdots & \vdots & \vdots & \ddots & \vdots \\ 1 & \lambda_{n+1} & \lambda_{n+1}^2 & \cdots & \lambda_{n+1}^n \end{pmatrix} \begin{pmatrix} a_0 \\ a_1 r \\ \vdots \\ a_n r^n \end{pmatrix} = \begin{pmatrix} 0 \\ 0 \\ \vdots \\ 0 \end{pmatrix}.$$

But a Vandermonde matrix has nonzero determinant, so it is invertible in the field of fractions of D. In particular, as D has no zero divisors in R, we may work in $D^{-1}R$, hence $a_n r^n = 0$. Since $0 \neq a_n \in D$, $r^n = 0$. □

Lemma 1.2.5. *There exists a positive integer n, determined by $w(x,y)$ and char F, so that every F-algebra R such that $\mathscr{U}(R)$ satisfies $w(x,y) = 1$ has the following property. Namely, if $a, b \in R$, $a^2 = b^2 = 0$, but $(ab)^n \neq 0$, then $|F| \leq n$ and ab is not nilpotent.*

Proof. Consider the polynomial $f(x)$ from Lemma 1.2.3, and let n be its degree. If $\lambda \in F$, then $(\lambda a)^2 = 0$, so $f(\lambda ab) = 0$. By Lemma 1.2.4, if F has more than n elements, then $(ab)^n = 0$, a contradiction.

Also, if $(ab)^m = 0$ for some m, then notice that the minimal polynomial for ab divides f and x^m, so it must be x^k for some $k \leq n$; again, we have a contradiction. □

Now we can give some restrictions on matrix rings whose unit groups satisfy $w(x,y) = 1$. As usual, we write $GL_n(F)$ for the unit group of $M_n(F)$, and E_{ij} for the matrix having a 1 in the i, j position and zeroes elsewhere.

Lemma 1.2.6. *There exists an integer m, determined by the group identity $w(x,y) = 1$ and char F, so that if $GL_k(F)$ (with $k \geq 2$) satisfies $w(x,y) = 1$, then $k \leq m$ and $|F| \leq m$.*

Proof. Notice that $E_{12}^2 = E_{21}^2 = 0$, but $E_{12}E_{21} = E_{11}$, which is not nilpotent. Thus, by Lemma 1.2.5, $|F| \leq n$, where n is as in the statement of that lemma.

Choose an extension field F' of F of smallest possible degree so that $|F'| > n$. Then regarding F' as an F-module, we note that F' acts on itself via multiplication, and so may be regarded as a unital subring of $\text{End}_F F'$. But, of course, choosing a basis for F' over F, we then identify $\text{End}_F F'$ with $M_r(F)$, where $|F|^r = |F'|$. That is, $M_2(F')$ is a unital subring of $M_{2r}(F)$. We already know that $GL_2(F')$ cannot satisfy $w(x,y) = 1$, hence $GL_{2r}(F)$ cannot. Letting m be the larger of n and $2r$, we have our result, subject to putting a bound on r, independent of the choice of F. But finite fields have extensions of every degree, so any $r \geq \log_p(n+1) \geq \log_{|F|}(n+1)$ will work, where char $F = p$. □

The polynomial identity situation for matrix rings is well understood. For any positive integer r, the standard polynomial identity of degree r is

$$s_r(x_1, \ldots, x_r) = \sum_{\sigma \in S_r} a_\sigma x_{\sigma(1)} x_{\sigma(2)} \cdots x_{\sigma(r)},$$

where the sum ranges over the symmetric group S_r, and a_σ is 1 if σ is even and -1 if σ is odd. The following is a classical result due to Amitsur and Levitzki.

Proposition 1.2.7. *For any positive integer* k, $M_k(F)$ *satisfies* $s_{2k}(x_1,\ldots,x_{2k}) = 0$.

Proof. See [82, Theorem 5.1.9].   □

We can now complete the proof for locally finite p'-groups. (By a p'-group we mean a group with no elements of order p.) If $\alpha \in FG$, then the support of α, $\mathrm{supp}(\alpha)$, is the set of all $g \in G$ with nonzero coefficients in α.

Proposition 1.2.8. *Let* F *be a field of characteristic* $p \geq 0$ *and* G *a locally finite group. If* $p > 0$, *suppose that* G *has no* p-*elements. If* $\mathscr{U}(FG)$ *satisfies a group identity, then* FG *satisfies a polynomial identity.*

Proof. Let us say that $\mathscr{U}(FG)$ satisfies $w(x,y) = 1$, and let m be the integer from Lemma 1.2.6. Choose any $\alpha_i \in FG$, $1 \leq i \leq 2m$, and let H be the (necessarily finite) subgroup of G generated by the supports of the α_i. Then $FH = \bigoplus_i M_{n_i}(D_i)$, where each D_i is an F-division algebra. Since $\mathscr{U}(FH)$ satisfies $w(x,y) = 1$, so does $GL_{n_i}(D_i)$. By Proposition 1.2.2, each D_i is therefore a field, and by Lemma 1.2.6, $n_i \leq m$. Thus, by Proposition 1.2.7, $M_{n_i}(D_i)$ satisfies $s_{2m}(x_1,\ldots,x_{2m}) = 0$, and therefore so does FH. That is,
$$s_{2m}(\alpha_1,\ldots,\alpha_{2m}) = 0.$$
But the α_i were arbitrary elements of FG, and we are done.   □

Let us extend the result to semiprime group rings. Recall that a ring is said to be semiprime if it has no nonzero nilpotent ideals. The semiprime group rings were classified in a classical result due to Passman. Note that for any group G, we write $\phi(G)$ for the FC-centre; that is, the subgroup of G consisting of elements with only finitely many conjugates. We say that G is an FC-group if $G = \phi(G)$. For any prime p, we write $\phi_p(G)$ for the subgroup generated by the p-elements in $\phi(G)$.

Proposition 1.2.9. *Let* G *be any group. If* F *is a field of characteristic zero, then* FG *is semiprime. If* $\mathrm{char}\, F = p \geq 2$, *then the following are equivalent:*

 (i) FG *is semiprime,*
 (ii) G *does not have a finite normal subgroup with order divisible by* p, *and*
(iii) $\phi_p(G) = 1$.

Proof. See [82, Theorems 4.2.12 and 4.2.13].   □

We need a couple of lemmas to begin the semiprime case.

Lemma 1.2.10. *Suppose* R *is a ring and whenever* $a,b,c \in R$ *satisfy* $a^2 = bc = 0$, *we have* $bac = 0$. *Then every idempotent in* R *is central.*

Proof. Let e be an idempotent, and take any $r \in R$. Then let $a = er(1-e)$, $b = e$ and $c = 1-e$. Since $(1-e)e = 0$, we have $a^2 = bc = 0$, hence
$$0 = bac = e^2 r(1-e)^2 = er(1-e).$$

Therefore, $er = ere$. Interchanging b and c, and using $a = (1-e)re$, we get $re = ere$, hence $re = er$, and e is central.   □

Lemma 1.2.11. *Suppose R is a semiprime ring, and whenever $a^2 = bc = 0$, for $a, b, c \in R$, we have $bac = 0$. Then for any $b, c, d \in R$ with d nilpotent and $bc = 0$, we have $bdc = 0$.*

Proof. We suppose that this is true for all d such that $d^n = 0$, and prove that it is true for all d such that $d^{n+1} = 0$. Notice that

$$(1-d)(1+d+d^2+\cdots+d^n) = 1,$$

hence

$$0 = bc = (b(1-d))((1+d+d^2+\cdots+d^n)c).$$

Also, for any $r \in R$, $(crb)^2 = 0$, and therefore, by hypothesis,

$$b(1-d)(crb)(1+d+d^2+\cdots+d^n)c = 0.$$

Now, if $i \geq 2$, then $(d^i)^n = 0$, hence by induction, $bd^i c = 0$. It follows that

$$b(1-d)crb(1+d)c = 0,$$

and as $bc = 0$, we can reduce this to $bdcrbdc = 0$. Thus, for any $r_1, r_2 \in R$, $r_1 bdcrbdcr_2 = 0$, and it follows that if I is the ideal generated by bdc, then $I^2 = 0$. Thus, since R is semiprime, $I = 0$, and therefore $bdc = 0$.   □

Let us see what happens when FG is a semiprime group ring satisfying the above conditions. The following lemma will be needed for groups with elements of infinite order as well. If char $F = p > 0$, then we write P for the set of p-elements of G and Q for the set of p'-elements (of finite order). When $p = 0$, we let $P = 1$, and $Q = T$, the set of torsion elements.

Lemma 1.2.12. *Let G be a group, and let F be a field of characteristic $p \geq 0$ such that FG is semiprime. Let us suppose that for all $\alpha, \beta, \gamma \in FG$ with $\alpha^2 = \beta\gamma = 0$ we have $\beta\alpha\gamma = 0$. If $\mathscr{U}(FG)$ satisfies a group identity, then P and Q are (normal) subgroups of G, and Q is abelian. Furthermore, every subgroup of P is normal in T and every subgroup of Q is normal in G.*

Proof. Suppose $p > 0$, and take $g, h \in P$. Then $(g-1)^{p^t} = 0$ for some $t \geq 0$. Also, $(1-h)\hat{h} = 0$. Thus, by Lemma 1.2.11, $(1-h)(g-1)\hat{h} = 0$. But as $(1-h)\hat{h} = 0$, this means that $(1-h)g\hat{h} = 0$, so $g\hat{h} = hg\hat{h}$. Since $g \in \text{supp}(g\hat{h})$, we have $g = hgh^i$ for some i. That is, $g^{-1}hg \in \langle h \rangle$, hence $\langle h \rangle$ is normal in $\langle P \rangle$, and similarly, so is $\langle g \rangle$. Thus, gh lies in $\langle g \rangle \langle h \rangle$, which is a subgroup of order dividing $o(g)o(h)$ and therefore a p-group. That is, the p-elements form a (normal) subgroup of G.

Take $g \in P$ and $h \in Q$. Then we still obtain $g = hgh^i$, hence $hg^{-1}h^{-1}g = h^{i+1}$. Now, $g \in P$, and P is a normal subgroup; hence the commutator $hg^{-1}h^{-1}g$ is in P, but h^{i+1} is a p'-element. It follows that h commutes with g. That is, $\langle g \rangle$ is normal in P and is centralized by every p'-element, so $\langle g \rangle$ is normalized by the torsion elements.

Now, let F have any characteristic and, take any $g \in Q$. Since char F does not divide $o(g)$, we have an idempotent $\frac{1}{o(g)}\hat{g}$. By Lemma 1.2.10, this idempotent is central. That is, $\langle g \rangle$ is normal in G, so every subgroup of Q is normal in G.

In particular, Q is locally finite, so consider any finite subgroup H of Q. Then FH is semisimple, and therefore a direct sum of matrix rings over division rings. As E_{11} is a noncentral idempotent in any 2×2 or larger matrix ring, FH is a direct sum of division rings. But in view of Proposition 1.2.2, each division ring must be a field. Thus H, and therefore Q, is abelian. □

Suppose that the conditions of the above lemma apply, and G is torsion. If $1 \neq g \in P$, then $\langle g \rangle$ is a finite normal p-subgroup, contradicting Proposition 1.2.9. We have the following.

Lemma 1.2.13. *Let F be a field and G a torsion group, such that FG is semiprime. If $\mathscr{U}(FG)$ satisfies a group identity, and if $\beta\alpha\gamma = 0$ for every $\alpha, \beta, \gamma \in FG$ with $\alpha^2 = \beta\gamma = 0$, then G is abelian.*

Thus, we are left with the situation where there exist $\alpha, \beta, \gamma \in FG$ with $\alpha^2 = \beta\gamma = 0$, but $\beta\alpha\gamma \neq 0$. Some additional terminology is required.

First, suppose that $0 \neq f(x_1, \ldots, x_n)$ is a polynomial identity for an F-algebra R (or, indeed, for any left or right ideal I in R). There is a standard linearization process, as follows. First, if any x_i does not appear in every monomial in f, then we substitute 0 for x_i and obtain a new polynomial identity in fewer variables. We repeat this until every variable appears in every monomial, and call this new polynomial identity f'.

Now, suppose that one of the variables, let us say x_1, appears two or more times in some monomial of $f'(x_1, \ldots, x_n)$. Then notice that I also satisfies

$$f''(x_1, \ldots, x_{n+1}) = f'(x_1 + x_{n+1}, x_2, \ldots, x_n) - f'(x_1, x_2, \ldots, x_n) - f'(x_{n+1}, x_2, \ldots, x_n).$$

Now, f'' is a polynomial of degree no greater than that of f', and in which every x_i appears in every monomial. Iterating this procedure, we eventually obtain a polynomial identity for I in which every variable appears exactly once in each monomial. We call such a polynomial multilinear. Thus, we have

Lemma 1.2.14. *Let F be a field and R an F-algebra. Let I be a right ideal in R. If I satisfies a polynomial identity of degree r, then I satisfies a multilinear polynomial identity of degree at most r.*

Let us discuss the notion of a generalized polynomial identity, or GPI. A GPI $f(x_1, \ldots, x_n)$ for an F-algebra R is a sum of terms of the form

$$r_0 x_{i_1} r_1 x_{i_2} \cdots r_k x_{i_{k+1}} r_{k+1},$$

where $r_i \in R$ and k is an integer, such that $f(a_1, \ldots, a_n) = 0$ for all $a_i \in R$. That is, elements of the ring are permitted to appear in the GPI, not just coefficients from the field. We can apply the same linearization process described above to a GPI, and obtain a multilinear GPI, which will be of the form

$$f(x_1,\ldots,x_n) = \sum_{\sigma \in S_n} f^\sigma(x_1,\ldots,x_n),$$

where each f^σ is a sum of terms of the form

$$r_0 x_{\sigma(1)} r_1 x_{\sigma(2)} \cdots r_{n-1} x_{\sigma(n)} r_n$$

for various $r_i \in R$.

Simply insisting that our GPI is nonzero is insufficient. Indeed, if r is central in R, then R satisfies $rx_1 - x_1 r$, but such an identity is not particularly helpful. Instead, we will insist upon nondegeneracy. A multilinear GPI is said to be nondegenerate if, for some $\sigma \in S_n$, f^σ is not a GPI for R. The group rings satisfying a nondegenerate multilinear GPI were classified by Passman.

Proposition 1.2.15. *Let F be a field and G a group. Then FG satisfies a nondegenerate multilinear GPI if and only if $(G : \phi(G)) < \infty$ and $(\phi(G))'$ is finite. In particular, if FG satisfies a polynomial identity, then $(G : \phi(G)) < \infty$ and $|(\phi(G))'| < \infty$.*

Proof. See [82, Theorem 5.3.15].   □

We will be able to make use of this repeatedly, beginning with

Lemma 1.2.16. *Let R be an F-algebra, and suppose that R contains a right ideal I such that I satisfies a polynomial identity of degree r, but $I^r \neq 0$. Then R satisfies a nondegenerate multilinear GPI.*

Proof. We may assume that I satisfies a multilinear polynomial identity

$$f(x_1,\ldots,x_n) = \sum_{\sigma \in S_n} f^\sigma(x_1,\ldots,x_n)$$

where

$$f^\sigma(x_1,\ldots,x_n) = a_\sigma x_{\sigma(1)} x_{\sigma(2)} \cdots x_{\sigma(n)},$$

$a_\sigma \in F$ and $n \leq r$. Relabeling if necessary, we assume that $a_{(1)} \neq 0$. Choosing $b_1,\ldots,b_n \in I$ with $b_1 \cdots b_n \neq 0$, we obtain a multilinear GPI for R, namely,

$$g(x_1,\ldots,x_n) = \sum_{\sigma \in S_n} g^\sigma(x_1,\ldots,x_n)$$

where

$$g^\sigma(x_1,\ldots,x_n) = a_\sigma b_{\sigma(1)} x_{\sigma(1)} \cdots b_{\sigma(n)} x_{\sigma(n)}.$$

We can see that this identity is nondegenerate by looking at the $\sigma = (1)$ term and plugging in $x_i = 1$ for all i.   □

We need a few more lemmas in order to complete the semiprime case.

Lemma 1.2.17. *Let R be a ring and I a nil ideal. If $\mathscr{U}(R)$ satisfies the group identity $w(x_1,\ldots,x_n) = 1$, then $\mathscr{U}(R/I)$ satisfies the same identity. Conversely, if R has prime characteristic p, I is nil of bounded exponent p^k, and $\mathscr{U}(R/I)$ satisfies the group identity $v(x_1,\ldots,x_n) = 1$, then $\mathscr{U}(R)$ satisfies $(v(x_1,\ldots,x_n))^{p^k} = 1$.*

Proof. Suppose $\mathscr{U}(R)$ satisfies $w(x,y) = 1$ and $\bar{r}_i \in \mathscr{U}(\bar{R}) = \mathscr{U}(R/I)$, $1 \leq i \leq n$. Then if $\bar{s}_i = (\bar{r}_i)^{-1}$, we can lift $\bar{r}_i$ and $\bar{s}_i$ up to r_i and s_i in R. If we let $u = r_i s_i - 1$, then $\bar{u} = 0$, so $u \in I$, and therefore $u^j = 0$ for some positive integer j. But then

$$r_i s_i (1 - u + u^2 - \cdots \pm u^{j-1}) = 1,$$

and similarly r_i has a left inverse. Thus, each $r_i \in \mathscr{U}(R)$, hence $w(r_1,\ldots,r_n) = 1$ and, therefore, $w(\bar{r}_1,\ldots,\bar{r}_n) = 1$, as required.

Conversely, if $r_1,\ldots,r_n \in \mathscr{U}(R)$, then $v(\bar{r}_1,\ldots,\bar{r}_n) = 1$, hence $v(r_1,\ldots,r_n) - 1 \in I$. But then $(v(r_1,\ldots,r_n) - 1)^{p^k} = 0$, and the result follows.   □

Lemma 1.2.18. *Let F be a field of characteristic $p > 0$ and G a group, such that $\mathscr{U}(FG)$ satisfies $w(x_1,\ldots,x_n) = 1$. If N is a normal p-subgroup of G, and either N is finite or G is locally finite, then $\mathscr{U}(F(G/N))$ satisfies $w(x_1,\ldots,x_n) = 1$.*

Proof. If N is finite, then by Lemma 1.1.1, $\Delta(G,N)$ is nilpotent, and the previous lemma does the job. So suppose that G is locally finite.

Take $\bar{\alpha}_1,\ldots,\bar{\alpha}_n \in \mathscr{U}(F(G/N))$. Lifting these elements up to $\alpha_1,\ldots,\alpha_n \in FG$, and similarly for their inverses, we let H be the subgroup generated by the supports of these elements. Then H is finite, hence $H \cap N$ is a finite normal p-subgroup of H. Now we can make use of the finite case. Since $\mathscr{U}(FH)$ satisfies $w(x_1,\ldots,x_n) = 1$, so does $\mathscr{U}(F(H/(H \cap N)))$. Replacing G with H, and N with $H \cap N$, we obtain our result.   □

Lemma 1.2.19. *Let G be a torsion group. If FG satisfies a nondegenerate multilinear GPI and $\mathscr{U}(FG)$ satisfies a group identity, then FG satisfies a polynomial identity.*

Proof. By Proposition 1.2.15, $(G : \phi(G)) < \infty$ and $|(\phi(G))'| < \infty$. Thus, G is an extension of a finite group by a torsion abelian group, and then by another finite group. Hence, G is locally finite. If $p = 0$, then Proposition 1.2.8 finishes the proof, so suppose p is prime. Let H be the centralizer of $(\phi(G))'$ in $\phi(G)$. As it is the intersection of the centralizers of finitely many elements, each such centralizer having finite index in $\phi(G)$, we have $(\phi(G) : H) < \infty$. Clearly $H' \leq (\phi(G))'$, so H' is central in H. That is, H is nilpotent. Thus, the p-elements of H form a normal subgroup of H. Let us call it N. By the previous lemma, $\mathscr{U}(F(H/N))$ satisfies $w(x,y) = 1$. Thus, by Proposition 1.2.8, $F(H/N)$ satisfies a polynomial identity. But then Proposition 1.1.4 tells us that H/N has a p-abelian (hence abelian) subgroup A/N of finite index. As H is of finite index in G, so is A. Also, A' is a subgroup of N, and N is a p-group. Since $A' \leq H' \leq (\phi(G))'$, A is p-abelian. Again by Proposition 1.1.4, this means that FG satisfies a polynomial identity.   □

Now we can deal with the semiprime case.

Proposition 1.2.20. *Let F be a field and G a torsion group, such that FG is semiprime. If $\mathscr{U}(FG)$ satisfies a group identity, then FG satisfies a polynomial identity.*

Proof. By Lemma 1.2.13, we may assume that there exist $\alpha, \beta, \gamma \in FG$ such that $\alpha^2 = \beta\gamma = 0$, but $\beta\alpha\gamma \neq 0$. Let $I = \beta\alpha\gamma FG$. If I is nilpotent, then clearly so is $(FG)I$, so since FG is semiprime, $I = 0$, and therefore $\beta\alpha\gamma = 0$, which is a contradiction. However, by Lemma 1.2.3, there is a nonzero polynomial $f(x) \in F[x]$ such that $f(\alpha\gamma\rho\beta) = 0$ for all $\rho \in FG$. Thus, $\beta f(\alpha\gamma\rho\beta)\alpha\gamma\rho = 0$ for all $\rho \in FG$. But this is a nonzero polynomial in $\beta\alpha\gamma\rho$. That is, $\beta\alpha\gamma FG$ satisfies a nonzero polynomial in $F[x]$. Therefore, by Lemma 1.2.16, FG satisfies a nondegenerate multilinear GPI, and Lemma 1.2.19 completes the proof. □

In particular, this proves Hartley's conjecture for the characteristic zero case. But we can prove a stronger statement there.

Corollary 1.2.21. *Let F be a field of characteristic zero and G a torsion group. Then $\mathscr{U}(FG)$ satisfies a group identity if and only if G is abelian.*

Proof. By Proposition 1.2.20, FG satisfies a polynomial identity, hence by Proposition 1.1.4, G is locally finite. Thus, it is enough to prove the result for finite groups G. But in this case, $FG = \bigoplus_i M_{n_i}(D_i)$ for division algebras D_i. By Proposition 1.2.2, each D_i is a field, and by Lemma 1.2.6, since F is infinite, each $n_i = 1$. Thus, FG is commutative. □

What if FG is not semiprime? Then it has a nonzero nilpotent ideal. For any ring R, let $N(R)$ denote the nilpotent radical of R; that is, the sum of all nilpotent ideals in R. Now, every finite sum of nilpotent ideals is nilpotent, but this is not necessarily true for infinite sums. Any element in the sum would appear in a finite sum of nilpotent ideals, so $N(R)$ must, at least, be nil. The following is a classical result due to Passman. (See [82, Theorem 8.1.12].)

Proposition 1.2.22. *Let F be a field of characteristic $p > 0$ and G a group. Then $N(FG)$ is nilpotent if and only if $\phi_p(G)$ is finite.*

We can now prove Hartley's conjecture when $N(FG)$ is nilpotent.

Lemma 1.2.23. *Let F be a field of characteristic $p \geq 2$ and G a torsion group. If $N(FG)$ is a nonzero nilpotent ideal and $\mathscr{U}(FG)$ satisfies a group identity, then FG satisfies a polynomial identity.*

Proof. Let H be the centralizer of $\phi_p(G)$ in G. Since $\phi_p(G)$ is finite, $(G : H) < \infty$. Now, if h is a p-element in $\phi_p(H)$, then h has finitely many conjugates in H. As H has finite index in G, there can be only finitely many conjugates in G as well. Thus, $\phi_p(H) \leq \phi_p(G)$, and therefore $N(FH)$ is nilpotent. Also, H centralizes $\phi_p(H)$, so $\phi_p(H)$ is abelian, hence, a finite p-group. If we can show that FH satisfies a polynomial identity, then H contains a p-abelian subgroup of finite index, and therefore, so does G. As $\mathscr{U}(FH)$ certainly satisfies a group identity, we may as well assume that $\phi_p(G)$ is a finite p-group.

Now, $\phi_p(G)$ is a characteristic subgroup of the normal subgroup $\phi(G)$, hence $\phi_p(G)$ is normal in G. By Lemma 1.2.18, $\mathscr{U}(F(G/\phi_p(G)))$ satisfies a group identity. If we can show that $F(G/\phi_p(G))$ satisfies a polynomial identity $f(x_1,\ldots,x_r)$, then $f(\alpha_1,\ldots,\alpha_r) \in \Delta(G,\phi_p(G))$ for all $\alpha_i \in FG$. By Lemma 1.1.1, $\Delta(G,\phi_p(G))$ is nilpotent, hence FG satisfies $(f(x_1,\ldots,x_r))^{p^k}$ for some k. Thus, we may factor out $\phi_p(G)$ and assume that $\phi_p(G) = 1$. But then by Proposition 1.2.9, FG is semiprime, so by Proposition 1.2.20, FG satisfies a polynomial identity, and we are done. $\square$

Finally, suppose that $N(FG)$ is not nilpotent. We need an easy ring-theoretic lemma.

Lemma 1.2.24. *Let R be a ring, and suppose that $N(R)$ is not nilpotent. Then for any positive integer r, there exists a nilpotent ideal I of R such that $I^r \neq 0$.*

Proof. If this is not the case, then choose $\alpha_1,\ldots,\alpha_r \in N(R)$. As each α_i lies in a sum of nilpotent ideals, there must exist finitely many nilpotent ideals, $I_1,\ldots,I_q$ such that $\alpha_i \in I_1 + \cdots + I_q$ for all i. But a sum of finitely many nilpotent ideals is nilpotent, hence $(I_1 + \cdots + I_q)^r = 0$, and therefore $\alpha_1 \cdots \alpha_r = 0$. That is, $N(R)$ is nilpotent, contradicting our assumption. $\square$

The following fact is well known.

Lemma 1.2.25. *Let $F\{x_1,x_2,\ldots\}$ be the free algebra on noncommuting indeterminates $x_1,x_2,\ldots$, and let $R = F\{x_1,x_2,\ldots\}[[z]]$ denote its power series ring. Then for any positive integer k, the elements $1 + x_i z$, $1 \leq i \leq k$, generate a free subgroup of $\mathscr{U}(R)$.*

Proof. Suppose that

$$(1+x_{i_1}z)^{r_1} \cdots (1+x_{i_l}z)^{r_l} = 1,$$

where $1 \leq i_j \leq k$, and each $r_j \neq 0$. If char $F = p$, write $r_j = p^{u_j} v_j$, where p does not divide v_j. (If $p = 0$, just take $p^{u_j} = 1$ for all j.) Then

$$1 = (1+y_{i_1}z^{p^{u_1}})^{v_1} \cdots (1+y_{i_l}z^{p^{u_l}})^{v_l},$$

where $y_{i_j} = x_{i_j}^{p^{u_j}}$ for all j. But noting that for any $s \in F\{x_1,x_2,\ldots\}$, we have

$$(1+sz)^m = 1 + msz + \cdots$$

for any integer m, we see that the coefficient of

$$y_{i_1} y_{i_2} \cdots y_{i_l} z^{p^{u_1}+p^{u_2}+\cdots+p^{u_l}}$$

is $v_1 v_2 \cdots v_l$. This, however, is not divisible by char F, hence it is not zero. We have a contradiction. $\square$

The last part of the proof is the following.

Lemma 1.2.26. *Let R be an F-algebra. Suppose that $\mathscr{U}(R)$ satisfies the group identity $w(x,y) = 1$ and $N(R)$ is not nilpotent. Then R satisfies a nondegenerate multilinear GPI.*

Proof. Let us work in the power series ring $F\{x_1,x_2\}[[z]]$, as in the preceding lemma. Since $1 + x_1 z$ and $1 + x_2 z$ generate a free group, we get

$$0 \neq w(1 + x_1 z, 1 + x_2 z) - 1.$$

Writing $(1 + x_j z)^{-1} = 1 - x_j z + \cdots$, and expanding the above expression, we get

$$0 \neq w(1 + x_1 z, 1 + x_2 z) - 1 = \sum_{i \geq 1} f_i(x_1, x_2) z^i,$$

where each f_i is a homogeneous polynomial of degree i. (That is, every monomial in f_i has degree i.)

Let r be the smallest integer such that f_r is not the zero polynomial. Find a nilpotent ideal I of R, as in Lemma 1.2.24, so that $I^s \neq 0$, but $I^{s+1} = 0$, where $s \geq r$. Choose any $\alpha_1, \alpha_2, \alpha_3 \in I$. Then as $\alpha_i^{s+1} = 0$, each $1 + \alpha_i$ is a unit, and therefore

$$0 = w(1 + \alpha_1, 1 + \alpha_2) - 1 = \sum_{i \geq r} f_i(\alpha_1, \alpha_2).$$

As $f_i(\alpha_1, \alpha_2)$ is a sum of products of i elements of I, and $I^{s+1} = 0$, we can write instead that

$$\sum_{i=r}^{s} f_i(\alpha_1, \alpha_2) = 0,$$

and therefore

$$\sum_{i=r}^{s} f_i(\alpha_1, \alpha_2) \alpha_3^{s-r} = 0.$$

Again, a product of $s + 1$ elements of I gives zero, so we are left with

$$f_r(\alpha_1, \alpha_2) \alpha_3^{s-r} = 0.$$

That is, $f_r(x_1, x_2) x_3^{s-r}$ is a polynomial identity of degree s for I. But $I^s \neq 0$, so by Lemma 1.2.16, R satisfies a nondegenerate multilinear GPI. $\square$

Combining Lemmas 1.2.19 and 1.2.26, we have our main result for this section, when $N(FG)$ is not nilpotent. The other cases having been dealt with in Proposition 1.2.20 and Lemma 1.2.23, we are done. The main result, from Giambruno et al. [31, 38] and Liu [75], is

Theorem 1.2.27. *Let F be a field and G a torsion group. If $\mathscr{U}(FG)$ satisfies a group identity, then FG satisfies a polynomial identity.*

1.3 Group Rings of Torsion Groups

A positive answer to Hartley's conjecture having been established, it was natural to look for necessary and sufficient conditions for $\mathscr{U}(FG)$ to satisfy a group identity. Clearly, satisfying a polynomial identity cannot be sufficient. We see from Proposition 1.1.4 that if G is finite, then FG always satisfies a polynomial identity, but if char $F = 0$, then by Corollary 1.2.21, $\mathscr{U}(FG)$ does not satisfy a group identity unless G is abelian. Thus, we will assume throughout this section that F is a field of characteristic $p > 0$.

The necessary and sufficient conditions in the case where G is a torsion group were determined for infinite fields by Passman in [84]. It turns out that the solution for finite fields is different if G' is not a p-group. In [76], Liu and Passman presented the solution for finite fields. The main results for this section are the following:

Theorem 1.3.1. *Let F be a field of characteristic $p > 0$ and G a torsion group. If G' is a p-group, then the following are equivalent:*

(i) $\mathscr{U}(FG)$ *satisfies a group identity;*
(ii) $\mathscr{U}(FG)$ *satisfies the group identity* $(x,y)^{p^r} = 1$, *for some positive integer r;*
(iii) G *has a p-abelian subgroup of finite index, and G' has bounded exponent.*

Theorem 1.3.2. *Let F be a field of characteristic $p > 0$ and G a torsion group. If G' is not a p-group, then the following are equivalent:*

(i) $\mathscr{U}(FG)$ *satisfies a group identity;*
(ii) $\mathscr{U}(FG)$ *has bounded exponent;*
(iii) G *has a p-abelian subgroup of finite index, G has bounded exponent, and F is a finite field.*

In particular, we can see that if the units of a group ring of a torsion group satisfy a group identity, then they satisfy a group identity having a particularly nice form.

For each theorem, it is obvious that (ii) implies (i). We will begin by showing that (i) implies (iii).

Recall that for a ring R, the Jacobson radical $J(R)$ is the intersection of all maximal left ideals of R. The Jacobson radical contains every nil ideal of R (see, for instance, [62, Proposition 10.27]). It is an important ideal in ring theory, and we record a few well-known results about $J(FG)$.

Proposition 1.3.3. *Let F be a field and G a group. Then $J(FG)$ has the following properties:*

1. *If G is finite, then $J(FG)$ is nilpotent, and $FG/J(FG)$ is a direct sum of matrix rings over F-division algebras.*
2. *If G is finite and F is a perfect field, then FG has an F-subalgebra isomorphic to $FG/J(FG)$.*
3. *If F' is an extension field of F, then $J(F'G) \cap FG \subseteq J(FG)$.*
4. *If F' is an algebraic extension field of F, then $J(FG) \subseteq J(F'G)$.*
5. *If H is a subgroup of G, then $J(FG) \cap FH \subseteq J(FH)$.*

6. If H is a normal subgroup of G of finite index, then $J(FH) \subseteq J(FG)$.

Proof. For the first part, we simply note that since FG is finite-dimensional, it is surely Artinian. Thus, by [62, Theorem 4.12], $J(FG)$ is nilpotent. Also, by [62, Proposition 4.6 and Theorem 4.14], $FG/J(FG)$ is semisimple, and the Wedderburn–Artin theorem applies. The second part is Wedderburn's principal theorem applied to FG (see [92, Theorem 2.5.37]). The remaining parts are Lemma 7.1.4, Theorem 7.2.13, Lemma 7.1.5 and Theorem 7.2.7, respectively, in [82]. □

We assume that our field F has characteristic $p > 0$, and begin by demonstrating that if $\mathscr{U}(FG)$ satisfies a group identity, then G' has bounded exponent. Due to Theorem 1.2.27 and Proposition 1.1.4, we can always assume that our group is locally finite. We begin with an easy lemma. As usual, we write $\alpha^g = g^{-1}\alpha g$ for any $g \in G$ and $\alpha \in FG$.

Lemma 1.3.4. *Let G be a group having an abelian normal subgroup A, such that $G/A = \langle Ag \rangle$. Then $G' = \{(a,g) : a \in A\}$.*

Proof. Define $\rho : A \to A$ via $\rho(a) = (a,g)$. Since A is abelian and normal, ρ is a homomorphism. Let H be its image. Evidently, $H \leq G'$. Also, H is centralized by A and $(a,g)^g = (a^g, g)$, so H is normal in G. But now for any $a \in A$ and integer i, we have $a^{g^i} = ah$ for some $h \in H$. That is, A/H is central in G/H. But $(G/H)/(A/H)$ is cyclic, so G/H is cyclic modulo its centre. Thus, G/H is abelian, so $H = G'$, as required. □

Let us establish some special cases.

Lemma 1.3.5. *Suppose that G is the semidirect product $A \rtimes \langle g \rangle$, where A is an abelian p-group and g has prime order $q \neq p$. If $\mathscr{U}(FG)$ satisfies a group identity, then G' has bounded exponent.*

Proof. Let us define a function $\theta : FA \to FA$ via

$$\theta(\alpha) = \alpha + \alpha^g + \alpha^{g^2} + \cdots + \alpha^{g^{q-1}}.$$

Since FA is commutative and g commutes with $\theta(\alpha)$, we see that $\theta(\alpha)$ is central in FG for all $\alpha \in FA$. Fix any $a \in A$, and let $\beta = \hat{g}a^{-1}(1 - g^{-1})$. As $(1 - g^{-1})\hat{g} = 0$, we have $\beta^2 = 0$. Also, let $\gamma = (qa - \theta(a))\hat{g}$. Now, $qa - \theta(a) \in \Delta(A)$, and considering finitely generated (hence finite) subgroups of A, we see from Lemma 1.1.1 that $\Delta(A)$ is nil. Thus, $\gamma \in J(FA)FG \subseteq J(FG)$ by Proposition 1.3.3. Take any $\delta \in J(FG)$. Then since G is locally finite, $\delta \in FH$ for some finite subgroup H. Thus, by Proposition 1.3.3, $\delta \in J(FH)$, and $J(FH)$ is nilpotent, hence δ is nilpotent. That is, $\beta\gamma \in J(FG)$ is a nilpotent element. But also, for any $\alpha \in FA$, we have

$$\hat{g}\alpha\hat{g} = \sum_{i=0}^{q-1}\sum_{j=0}^{q-1} g^i \alpha g^j = \theta(\alpha)\hat{g}.$$

Thus,

$$\gamma^2 = (qa - \theta(a))\hat{g}(qa - \theta(a))\hat{g} = (qa - \theta(a))\theta(qa - \theta(a))\hat{g}.$$

Since $\theta(a)$ is central, we get $\theta(\theta(a)) = q\theta(a)$, and therefore $\gamma^2 = 0$.

By Lemma 1.2.5, there exists a positive integer n, depending only upon our group identity, such that $(\beta\gamma)^n = 0$. Choose r so that $p^r \geq n$. Then $(\beta\gamma)^{p^r} = 0$. Also,

$$\beta\gamma = \hat{g}a^{-1}(1 - g^{-1})(qa - \theta(a))\hat{g} = \hat{g}a^{-1}(1 - g^{-1})qa\hat{g},$$

since $\theta(a)$ is central and $(1 - g^{-1})\hat{g} = 0$. Therefore,

$$\beta\gamma = q\hat{g}(1 - a^{-1}g^{-1}ag)\hat{g}.$$

Of course, $(\hat{g})^2 = q\hat{g}$, so

$$\beta\gamma = q(q - \hat{g}(a,g))\hat{g}.$$

But we have seen that $\hat{g}(a,g)\hat{g} = \theta((a,g))\hat{g}$, so

$$\beta\gamma = q(q - \theta((a,g)))\hat{g}.$$

Notice that for any $\alpha \in FA$, we have

$$(\theta(\alpha))^{p^r} = \alpha^{p^r} + (\alpha^g)^{p^r} + \cdots + (\alpha^{g^{q-1}})^{p^r},$$

since each α^{g^i} lies in FA, which is commutative. That is, $\theta(\alpha^{p^r}) = (\theta(\alpha))^{p^r}$. Hence,

$$0 = (\beta\gamma)^{p^r} = q^{p^r}(q^{p^r} - \theta((a,g)^{p^r}))q^{p^r-1}\hat{g},$$

since $\theta((a,g))$ is central, and $(\hat{g})^m = q^{m-1}\hat{g}$ for any positive integer m. Thus, since $q^{p^r} = q$, we have

$$(q - \theta((a,g)^{p^r}))\hat{g} = 0.$$

Notice, however, that $q - \theta((a,g)^{p^r}) \in FA$, and elements of FG are uniquely expressed in the form $\sum_{i=0}^{q-1} \alpha_i g^i$, with $\alpha_i \in FA$. In particular,

$$\theta((a,g)^{p^r}) = q.$$

But $(a,g)^{p^r}$ is simply an element of A, and applying θ, we obtain a sum of its conjugates. The only way 1 can appear as one of those group elements is if $(a,g)^{p^r} = 1$. By the previous lemma, G' has bounded exponent. $\qquad\square$

Lemma 1.3.6. *Suppose $G = A \rtimes \langle g \rangle$, where A is an abelian p-group, and $o(g) = p$. If $\mathscr{U}(FG)$ satisfies a group identity, then G' has bounded exponent.*

Proof. Define θ as in the previous proof (with $q = p$), and fix $a \in A$. Notice that $(\hat{g})^2 = (a^{-1}\hat{g}a)^2 = 0$. Also, $\hat{g}$ has augmentation zero, so $\beta = a^{-1}\hat{g}a\hat{g} \in \Delta(G)$. But G is a locally finite p-group, so Lemma 1.1.1 tells us that β is nilpotent. Thus, by Lemma 1.2.5, there exists an n, depending only upon the group identity, such that $\beta^n = 0$. Choosing r so that $p^r \geq n$, we have $\beta^{p^r} = 0$.

We know that for any $\alpha \in FA$, $\theta(\alpha)$ is central and $\hat{g}\alpha\hat{g} = \theta(\alpha)\hat{g}$. Thus,

$$\beta^2 = a^{-1}\hat{g}a(\hat{g}a^{-1}\hat{g})a\hat{g}$$
$$= a^{-1}\hat{g}a\theta(a^{-1})\hat{g}a\hat{g}$$
$$= \theta(a^{-1})a^{-1}(\hat{g}a\hat{g})a\hat{g}$$
$$= \theta(a^{-1})a^{-1}\theta(a)\hat{g}a\hat{g}$$
$$= \theta(a^{-1})\theta(a)\beta.$$

It follows that

$$0 = \beta^{p^r} = (\theta(a^{-1})\theta(a))^{p^r-1}\beta$$

and in the same manner,

$$0 = \hat{g}\beta^{p^r} = (\theta(a^{-1}))^{p^r}(\theta(a))^{p^r}\hat{g}.$$

Now, $\theta(a)$ and $\theta(a^{-1})$ lie in FA, so the only way this can happen is if

$$(\theta(a^{-1}))^{p^r}(\theta(a))^{p^r} = 0.$$

But we already know that $(\theta(a^{-1}))^{p^r}(\theta(a))^{p^r} = \theta(a^{-p^r})\theta(a^{p^r})$. Let $b = a^{p^r}$. Then

$$0 = \theta(b^{-1})\theta(b)$$
$$= (b^{-1} + (b^{-1})^g + \cdots + (b^{-1})^{g^{p-1}})(b + b^g + \cdots + b^{g^{p-1}})$$
$$= \sum_{i=0}^{p-1} \theta(b^{-1}b^{g^i}).$$

However, the $i = 0$ term here is $\theta(1) = p = 0$, so in fact,

$$\sum_{i=1}^{p-1} \theta((b, g^i)) = 0.$$

When we apply θ to a group element, we obtain a sum of p group elements, so the left-hand side of this last equation is a sum of $p(p-1)$ group elements. If we are to get zero, then they must cancel each other in sets of size p. That is, there can be at most $p-1$ distinct group elements appearing. But the conjugates of a group element by the powers of an element of prime order are either all equal or all distinct. Thus, if (b, g) is not central, then it will produce p distinct conjugates in the expression above. This is impossible, so (b, g) is central. Thus, $(b, g)^{g^i} = (b, g)$ for all i, and

$$(b, g)^p = b^{-1}b^g(b^{-1}b^g)^g \cdots (b^{-1}b^g)^{g^{p-1}}$$
$$= b^{-1}g^{-1}bgg^{-1}b^{-1}g^{-1}bg^2 \cdots g^{p-1}$$
$$= b^{-1}g^{-p}bg^p = 1.$$

Since the function given by $a \mapsto (a, g)$ is a homomorphism on A, we have

$$1 = (b,g)^p = (a^{p^r},g)^p = (a,g)^{p^{r+1}}$$

for all $a \in A$. By Lemma 1.3.4, this means that G' has bounded exponent. $\square$

In order to generalize this a bit, we need a famous group-theoretic result due to Schur. Write $\zeta(G)$ for the centre of G.

Proposition 1.3.7. *Let G be any group.*

1. *If $(G : \zeta(G)) = n < \infty$, then G' is finite and the exponent of G' divides n.*
2. *If G is solvable and, for some prime p, $G/\zeta(G)$ has exponent p^k, then G' has exponent p^m for some m.*

Proof. See [94, Theorem I.4.2 and Corollary I.4.3]. $\square$

The following fact will be useful later as well.

Lemma 1.3.8. *Let F be any field and G any group. Let N be a finite normal subgroup of G. Then, for $\alpha \in FG$, we have $\alpha\hat{N} = 0$ if and only if $\alpha \in \Delta(G,N)$.*

Proof. Let X be a transversal of N in G. Then we can write $\alpha = \alpha_1 g_1 + \cdots + \alpha_n g_n$, with each $\alpha_i \in FN$ and the g_i distinct elements of X. As $\hat{N}$ is central, we have $\alpha\hat{N} = \alpha_1\hat{N}g_1 + \cdots + \alpha_n\hat{N}g_n$. It follows that $\alpha\hat{N} = 0$ if and only if $\alpha_i\hat{N} = 0$ for each i. But $\alpha_i\hat{N} = \varepsilon(\alpha_i)\hat{N}$, so each $\alpha_i\hat{N} = 0$ if and only if each $\alpha_i \in \Delta(N)$. This is equivalent to saying that $\alpha \in \Delta(G,N)$. $\square$

We can use this to prove the analogue of Lemma 1.2.18 for p'-subgroups.

Lemma 1.3.9. *Let G be a group and F a field. Let N be a torsion normal subgroup of G containing no elements of order divisible by the characteristic of F. If N is finite or G is locally finite, and $\mathcal{U}(FG)$ satisfies $w(x_1,\ldots,x_n) = 1$, then $\mathcal{U}(F(G/N))$ satisfies $w(x_1,\ldots,x_n) = 1$.*

Proof. Suppose N is finite. Let $e = \frac{\hat{N}}{|N|}$. Then e is a central idempotent, and $FG = FGe \oplus FG(1-e)$. Thus, $\mathcal{U}(FG) = \mathcal{U}(FGe) \times \mathcal{U}(FG(1-e))$. But notice that the kernel of $\theta : FG \to FGe$, given by $\theta(\alpha) = \alpha e$, is $\Delta(G,N)$, by the last lemma. Thus, $F(G/N) \cong FG/\Delta(G,N) \cong FGe$. Therefore, $\mathcal{U}(F(G/N))$ is isomorphic to a subgroup of $\mathcal{U}(FG)$.

If G is locally finite, then we may select $\bar{\alpha}_1,\ldots,\bar{\alpha}_n \in \mathcal{U}(F\bar{G}) = \mathcal{U}(F(G/N))$ and lift them to $\alpha_1,\ldots,\alpha_n \in FG$, and similarly for their inverses. Then we can take a finitely generated (hence finite) subgroup H of G containing the supports of these elements of FG, and we see that $\mathcal{U}(F(H/(N \cap H)))$ satisfies $w(x_1,\ldots,x_n) = 1$. Thus, $w(\bar{\alpha}_1,\ldots,\bar{\alpha}_n) = 1$, as required. $\square$

Now we can prove

Lemma 1.3.10. *Suppose G contains an abelian normal p-subgroup A of finite index. If $\mathcal{U}(FG)$ satisfies a group identity, then G' has bounded exponent.*

Proof. We first consider the case in which G/A is cyclic, say $G/A = \langle Ag \rangle$. Our proof is by induction on $o(g)$. If this order is prime, then Lemmas 1.3.5 and 1.3.6 give us the result. So let $o(g)$ be composite, and let q be a prime dividing $o(g)$. Let $H = \langle A, g^q \rangle$. By induction, H' has bounded exponent. Also, by Lemma 1.3.4, $H' = \{(a, g^q) : a \in A\}$. Clearly, $N = \langle H', g^q \rangle$ has bounded exponent as well.

Now, H' is certainly centralized by A and normalized by g, so H' is normal in G, and as $H' \leq N$, it is clear that the conjugates of g^q lie in N as well. Thus, N is normal in G. But now G/N is generated by AN/N and Ng. Furthermore, since G is cyclic modulo a p-group, its p-elements form a subgroup. In particular, the p-elements of N form a subgroup, N_1. Surely G is locally finite hence, by Lemma 1.2.18, $\mathscr{U}(F(G/N_1))$ satisfies a group identity. But the preceding lemma tells us that we can factor out N/N_1 as well. Thus, $\mathscr{U}(F(G/N))$ satisfies a group identity. Since $o(Ng) \leq q < o(g)$, it follows by induction that $(G/N)'$ has bounded exponent. But $(G/N)' = G'N/N$, and as N has bounded exponent, so does G'.

Let us deal with the general case. Consider all subgroups H_i of G containing A such that H_i/A is cyclic. As G/A is finite, there are only finitely many such subgroups. We know that each H_i' has bounded exponent, so as all H_i' lie inside A, which is abelian, the subgroup K generated by all of the H_i' has bounded exponent. Of course, every conjugate of an H_i is an H_j for some j, and similarly for H_i'. Thus, K is normal. Also, A/K is central in G/K. Indeed, if $a \in A$, $g \in G$, then letting $H_i = \langle A, g \rangle$, we have $(a, g) \in H_i' \leq K$. Thus, the centre of G/K has finite index, so by Proposition 1.3.7, $(G/K)'$ is finite. But $(G/K)' = G'K/K$. Since K has bounded exponent, so does G'. $\qquad\square$

Let us prove that (i) implies (iii) in Theorem 1.3.1.

Proposition 1.3.11. *Let G be a torsion group and F a field of characteristic $p > 0$. If $\mathscr{U}(FG)$ satisfies a group identity, and G' is a p-group, then G has a p-abelian subgroup of finite index, and G' has bounded exponent.*

Proof. By Theorem 1.2.27, FG satisfies a polynomial identity, so by Proposition 1.1.4, G has a p-abelian subgroup A of finite index. Taking the core of A in G, we may as well assume that A is normal. By Lemma 1.2.18, $\mathscr{U}(F(G/A'))$ satisfies a group identity as well, and if we can show that $(G/A')'$ has bounded exponent, then since $(G/A')' = G'A'/A'$, and A' is finite, it will follow that G' has bounded exponent. Thus, we can work with G/A' or, effectively, assume that A is abelian.

Let us write $A = H \times K$, where H is a p-group and K is a p'-group. Since K is characteristic in A, it is normal in G. By Lemma 1.3.9, $\mathscr{U}(F(G/K))$ satisfies a group identity, as G is locally finite. If we can show that $(G/K)'$ has bounded exponent, then $G'K/K$ has bounded exponent. But G' is a p-group, hence $G' \cap K = 1$. Therefore, $G' \simeq G'/(G' \cap K)$ has bounded exponent. Thus, we may assume that A is a p-group. Lemma 1.3.10 completes the proof. $\qquad\square$

To complete the proof of Theorem 1.3.1, we must show that (iii) implies (ii). We will need the following well-known lemma.

Lemma 1.3.12. *Let G be a group, and H a subgroup of index $n < \infty$. Let $g_1, \ldots, g_n$ be a right transversal of H in G. For any field F and any $\alpha \in FG$, write $g_i \alpha = \sum_{j=1}^{n} \alpha_{ij} g_j$, with $\alpha_{ij} \in FH$. Then the map $\theta : FG \to M_n(FH)$ given by $\theta(\alpha) = (\alpha_{ij})$ is an F-algebra embedding.*

Proof. Elements of FG are uniquely written in the form $\sum_{i=1}^{n} \alpha_i g_i$, with $\alpha_i \in FH$. That is, FG is a free left FH-module with basis $\{g_1, \ldots, g_n\}$. But FG is a right FG-module as well. For each $\alpha \in FG$, define $\rho_\alpha : FG \to FG$ via $\rho_\alpha(\beta) = \beta\alpha$. Then for any $\gamma \in FH$, we have

$$\rho_\alpha(\gamma\beta) = \gamma\beta\alpha = \gamma\rho_\alpha(\beta)$$

for all $\beta \in FG$. That is, ρ_α is a homomorphism of left FH-modules. Furthermore, if $\rho_{\alpha_1} = \rho_{\alpha_2}$, then $\rho_{\alpha_1}(1) = \rho_{\alpha_2}(1)$, hence $\alpha_1 = \alpha_2$. Thus, the map $\alpha \mapsto \rho_\alpha$ embeds FG in the ring of left FH-module homomorphisms from FG to FG. Since FG is free over FH, we identify FG with a subring of $M_n(FH)$. In fact, the matrix of ρ_α is determined by its action on the basis, $\{g_1, \ldots, g_n\}$. But this is the matrix defined in the statement of the lemma.   □

Lemma 1.3.13. *Let G be a group and A an abelian normal subgroup of index $n < \infty$. Suppose char $F = p > 0$ and I is an ideal of FA such that $g^{-1}\alpha g \in I$ for all $\alpha \in I$ and all $g \in G$. If I is nil of bounded exponent at most p^k, then $I(FG)$ is a nil ideal of FG of bounded exponent at most np^k.*

Proof. Let $g_1, \ldots, g_n$ be a transversal of A in G. If $\beta \in I(FG)$, then $g_i \beta = \beta^{g_i^{-1}} g_i \in I(FG)$, hence $g_i \beta = \sum_{j=1}^{n} \beta_{ij} g_j$, with each $\beta_{ij} \in I$. Thus, in the previous lemma, $\theta(I(FG)) \subseteq M_n(I)$. Therefore, it suffices to show that $M_n(I)$ is nil of the desired exponent.

Take any $M \in M_n(I)$, and let $f(x)$ be its characteristic polynomial. By the Cayley–Hamilton theorem, $f(M) = 0$. But this means that

$$M^n = \lambda_0 I_n + \lambda_1 M + \cdots + \lambda_{n-1} M^{n-1}$$

for some $\lambda_0, \ldots, \lambda_{n-1} \in I$, where I_n is the identity matrix. However, I is an ideal in the commutative ring FA, and the powers of M certainly commute; hence

$$M^{np^k} = \lambda_0^{p^k} I_n + \lambda_1^{p^k} M^{p^k} + \cdots + \lambda_{n-1}^{p^k} M^{(n-1)p^k} = 0,$$

as I has bounded exponent at most p^k. We are done.   □

The following observation essentially completes the proof of Theorem 1.3.1, and will be handy in the sequel as well.

Lemma 1.3.14. *Let F be a field of characteristic $p > 0$ and G any group such that FG satisfies a polynomial identity. If N is a normal p-subgroup of bounded exponent, then $\Delta(G, N)$ is nil of bounded exponent.*

Proof. We know from Proposition 1.1.4 that G has a p-abelian normal subgroup A of finite index. As $\Delta(G, A')$ is nilpotent, by Lemma 1.1.1, it suffices to work in $F(G/A')$ and therefore assume that A is abelian. Since $N/(N \cap A)$ is a finite p-group, we have $(\Delta(N/(N \cap A)))^{p^r} = 0$ for some r. That is, $(\Delta(N))^{p^r} \subseteq \Delta(N, N \cap A)$. Since N and $N \cap A$ are normal subgroups of G, it follows immediately that $(FG\Delta(N))^{p^r} \subseteq FG\Delta(N, N \cap A)$. That is, $(\Delta(G, N))^{p^r} \subseteq FG\Delta(N, N \cap A)$. Therefore, it suffices to show that $FG\Delta(N, N \cap A)$ is nil of bounded exponent. But $FG\Delta(N, N \cap A) = FG\Delta(N \cap A) = FG\Delta(A, N \cap A)$. By Lemma 1.3.13, it is enough to show that $\Delta(A, N \cap A)$ is nil of bounded exponent. However, A is abelian and $N \cap A$ is a p-group of bounded exponent. Thus, $\Delta(A, N \cap A)$ is nil of bounded exponent, as required.   □

We now present the

Proof of Theorem 1.3.1. As Proposition 1.3.11 gives us that (i) implies (iii), it remains only to verify that (iii) implies (ii). By Proposition 1.1.4, FG satisfies a polynomial identity. Thus, by Lemma 1.3.14, $\Delta(G, G')$ is nil of bounded exponent. By Lemma 1.2.17, it suffices to consider the group G/G'. But $\mathscr{U}(F(G/G'))$ satisfies $(x, y) = 1$, and we are done.   □

Let us now demonstrate that (i) implies (iii) in Theorem 1.3.2. We certainly know that G has a p-abelian subgroup of finite index. Let us show that the field must be finite.

Lemma 1.3.15. *If G is torsion, $\mathscr{U}(FG)$ satisfies a group identity, and G' is not a p-group, then F is a finite field.*

Proof. Assume that F is infinite. Since G is locally finite, we can find a finite subgroup H of G such that H' contains a p'-element. By Proposition 1.3.3, $J(FH)$ is nilpotent and $FH/J(FH) \cong \bigoplus_i M_{n_i}(D_i)$ is a direct sum of matrix rings over division rings. Hence, by Lemma 1.2.17, $\mathscr{U}(FH/J(FH))$ satisfies a group identity and therefore, so does each $GL_{n_i}(D_i)$. Now, by Proposition 1.2.2, each D_i is a field, and by Lemma 1.2.6, each $n_i = 1$. That is, $FH/J(FH)$ is commutative, and therefore, $h \in 1 + J(FH)$ for all $h \in H'$. Hence, $h - 1$ is nilpotent, and therefore $h^{p^k} = 1$ for some k. It follows that H' is a p-group, contradicting the choice of H.   □

It remains to show that G has bounded exponent. Another well-known fact is needed. If R is any ring, then we define the Lie product via $[r, s] = rs - sr$. We let $[R, R]$ be the additive subgroup of R generated by all $[r, s]$, $r, s \in R$.

Lemma 1.3.16. *If R is a ring with prime characteristic p, then for any $r_1, \ldots, r_n \in R$ and any positive integer m, we have $(\sum_{i=1}^n r_i)^{p^m} \equiv \sum_{i=1}^n r_i^{p^m} \pmod{[R, R]}$.*

Proof. Suppose that $m = 1$. By induction, it clearly suffices to show that $(r + s)^p \equiv r^p + s^p \pmod{[R, R]}$, for all $r, s \in R$. However,

$$(r + s)^p = r^p + s^p + \sum q_1 q_2 \cdots q_p,$$

where this sum extends over the $2^p - 2$ products with each $q_i \in \{r, s\}$, but not all of the q_i equal. But notice that

$$q_1q_2\cdots q_p - q_2q_3\cdots q_pq_1 = [q_1, q_2q_3\cdots q_p] \in [R,R]$$

and similarly for all other cyclic permutations of the symbols in $q_1\cdots q_p$. There are p such cyclic permutations (including the identity), so their sum is congruent modulo $[R,R]$ to $pq_1\cdots q_p = 0$. That is, $(r+s)^p \equiv r^p + s^p \pmod{[R,R]}$.

For the $m = 2$ case, we write $(r+s)^{p^2} = ((r+s)^p)^p = (r^p + s^p + d)^p$, where $d \in [R,R]$. Applying the $m = 1$ case again, we get that $(r+s)^{p^2} \equiv r^p + s^p + d^p$ $\pmod{[R,R]}$. If we can show that $d^p \in [R,R]$, then the $m = 2$ case is done, and the remaining cases will follow by induction. But if $a, b \in R$, then

$$[a,b]^p \equiv (ab)^p - (ba)^p \equiv [a, (ba)^{p-1}b] \equiv 0 \pmod{[R,R]}.$$

As $[R,R]$ is the set of sums of these $[a,b]$, the $m = 1$ case completes the proof. $\quad\square$

If $\alpha = \sum_{g\in G}\alpha_g g \in FG$, then the trace of α is $\mathrm{tr}(\alpha) = \alpha_1$.

Lemma 1.3.17. *Let G be a finite group and F a field of characteristic $p > 0$. If $\alpha \in FG$ is nilpotent, and α has no nontrivial p-elements in its support, then α has trace zero.*

Proof. Let $\alpha = \sum_{g\in G}\alpha_g g$. Choosing n so that $\alpha^{p^n} = 0$, the last lemma tells us that $\sum_{g\in G}\alpha_g^{p^n} g^{p^n} \in [FG, FG]$. Evidently, every element of $[FG, FG]$ has trace zero, as it consists of linear combinations of the terms $gh - hg$, with $g, h \in G$. But by our assumption on α, if $\alpha_g \neq 0$, then g^{p^n} cannot be the identity unless $g = 1$. Thus, $\mathrm{tr}(\sum_{g\in G}\alpha_g^{p^n} g^{p^n}) = \alpha_1^{p^n}$. Hence, α has trace zero. $\quad\square$

Lemma 1.3.18. *If G is torsion, $\mathscr{U}(FG)$ satisfies a group identity, and G' is not a p-group, then the p'-elements of G have bounded exponent.*

Proof. We know that G has a normal p-abelian subgroup A of finite index. By Lemma 1.2.18, $\mathscr{U}(F(G/A'))$ also satisfies a group identity, so it suffices to assume that A is abelian, and write $A = N \times K$, where N is a p'-group and K a p-group. Since A is of finite index, it is sufficient to bound the orders of the elements of N.

Fixing our group identity, we take the value m from Lemma 1.2.6. For all i from 2 to m, and all fields F' of characteristic p with $|F'| \leq m$, we consider the order of $GL_i(F')$. Let n be the least common multiple of all of these orders. We claim that if $h \in N$, then $h^{2n} = 1$. Suppose not.

By assumption, there exists $1 \neq g \in G'$ with g a p'-element. If $h^n = g^{-1}$, then as $o(h^n) > 2$, use g^{-1} in place of g. Then $1 \notin \{h^n, g, gh^n\}$. We may restrict ourselves to a finitely generated (hence finite) subgroup H of G, containing h, so that $g \in H'$. Then by Lemma 1.3.17, since $\alpha = (1-g)(1-h^n)$ does not have trace zero, if it is nilpotent, it must have p-elements in its support. Since N is normal and $h \in N$, gh^n is a p'-element. Thus, α is not nilpotent.

Now, since $J(FH)$ is nilpotent (by Proposition 1.3.3), $\alpha \notin J(FH)$. We know that $FH/J(FH) \cong \bigoplus_i M_{n_i}(D_i)$, where each D_i is a division ring. As each D_i must be finite, it is a field. Since $\alpha \notin J(FH)$, if we let $\theta_i : FH \to M_{n_i}(D_i)$ be the obvious map, then for some i, $\theta_i(\alpha) \neq 0$. If $n_i = 1$, then $M_{n_i}(D_i)$ is commutative, so clearly every commutator maps to 1, and $\theta_i(\alpha) = 0$. Now, as $J(FH)$ is nilpotent, $\mathscr{U}(FH/J(FH))$ satisfies our group identity, by Lemma 1.2.17. By Lemma 1.2.6, if $n_i \geq 2$, then $n_i \leq m$ and $|D_i| \leq m$. Since $\theta_i(h) \in GL_{n_i}(D_i)$, we must have $(\theta_i(h))^n = 1$, by choice of n. Therefore, $\theta_i(\alpha) = 0$, giving us a contradiction. $\qquad\square$

We need one more easy fact in order to complete the proof.

Lemma 1.3.19. *Let F be any field and G and H any groups. Then $F(G \times H)$ is isomorphic to $FG \otimes_F FH$.*

Proof. Evidently $(g, h) \mapsto g \otimes_F h$ is a group homomorphism from $G \times H$ into $\mathscr{U}(FG \otimes_F FH)$. Thus, we can extend it linearly to a ring homomorphism from $F(G \times H)$ to $FG \otimes_F FH$. Since the map sends a basis to a basis, it is an isomorphism. $\qquad\square$

We can now show that (i) implies (iii) in Theorem 1.3.2.

Proposition 1.3.20. *Let F be a field of characteristic $p > 0$ and G a torsion group. If $\mathscr{U}(FG)$ satisfies a group identity, and G' is not a p-group, then F is finite, G has a p-abelian subgroup of finite index, and G has bounded exponent.*

Proof. We know everything except that G has bounded exponent. Let A be the p-abelian subgroup of finite index, which we take to be normal. We have seen that we can work in $F(G/A')$, and therefore assume that A is abelian. Let $A = N \times K$, where N is a p'-group and K a p-group. Evidently, N is normal in G, and by Lemma 1.3.9, $\mathscr{U}(F(G/N))$ satisfies a group identity. But A/N is an abelian p-subgroup of finite index in G/N, hence $(G/N)' = G'N/N$ has bounded exponent, by Lemma 1.3.10. Since N has bounded exponent by Lemma 1.3.18, we see that G' has bounded exponent.

Also, it is sufficient to bound the exponent of A. Again, since N has bounded exponent, this simply means bounding the exponent of K. Now, $(G, K) \leq K \cap G'$, so (G, K) is a p-group of bounded exponent. By Lemma 1.2.18, $\mathscr{U}(F(G/(G, K)))$ satisfies a group identity, and surely $(G/(G, K))'$ is not a p-group. Thus, it suffices to show that $G/(G, K)$ has bounded exponent. That is, we may assume that K is central in G.

Take any $h \in K$, and choose $g \in G'$ so that g is a p'-element. As G is locally finite, there is a finite subgroup H of G containing h, so that $g \in H'$. Now, $H \cap K$ is a central p-subgroup and again, it suffices to work in $F(H/(H \cap K))$ and assume that $H \cap A$ is a p'-group. Since our field is finite, Proposition 1.3.3 tells us that FH contains an isomorphic copy of $FH/J(FH) \cong \bigoplus_i M_{n_i}(F_i)$, where each F_i is a finite division ring (hence, a field). If each $n_i = 1$, then $FH/J(FH)$ is commutative, hence every commutator in H lies in $1 + J(FH)$. But then $g - 1 \in J(FH)$, hence $g - 1$ is nilpotent, by Proposition 1.3.3. Thus, g must be a p-element, and we have a contradiction. Therefore, FH contains an isomorphic copy of $M_{n_i}(F_i)$, $n_i \geq 2$, and therefore of $M_2(F)$.

Now, $H \cap K = 1$, so by the previous lemma, $F(H \times K) \cong FH \otimes_F FK$, hence FG contains an isomorphic copy of $M_2(F) \otimes_F FK \cong M_2(FK)$. Notice that

$$\begin{pmatrix} 0 & h-1 \\ 0 & 0 \end{pmatrix}^2 = \begin{pmatrix} 0 & 0 \\ 1 & 0 \end{pmatrix}^2 = \begin{pmatrix} 0 & 0 \\ 0 & 0 \end{pmatrix}$$

and, furthermore,

$$\begin{pmatrix} 0 & h-1 \\ 0 & 0 \end{pmatrix} \begin{pmatrix} 0 & 0 \\ 1 & 0 \end{pmatrix} = \begin{pmatrix} h-1 & 0 \\ 0 & 0 \end{pmatrix}$$

is nilpotent, since h is a p-element. Thus, by Lemma 1.2.5, there exists an n depending only upon the group identity such that $(h-1)^n = 0$. Choosing r so that $p^r \geq n$, this means that $h^{p^r} = 1$, and we have a bound for the exponent of K. $\qquad \square$

Finally, we can complete the proof of Theorem 1.3.2. The fact that (iii) implies (ii) was originally proved in Coelho [25].

Proof of Theorem 1.3.2. We have only to prove that (iii) implies (ii). Let A be a p-abelian normal subgroup of finite index. By Lemmas 1.1.1 and 1.2.17, it suffices to show that $\mathscr{U}(F(G/A'))$ has bounded exponent. Thus, we will assume that A is abelian, and write $A = N \times K$, where N is a p'-group and K is a p-group. Since A is abelian and K has bounded exponent, it is clear that $\Delta(A,K)$ is nil of bounded exponent. As A and K are both normal in G, we can see from Lemma 1.3.13 that $\Delta(G,K)$ is also nil of bounded exponent. Thus, by Lemma 1.2.17, it suffices to work in $FG/\Delta(G,K) \cong F(G/K)$. That is, we may assume that A is a p'-group.

By Lemma 1.3.12, FG is embedded in $M_n(FA)$, where $n = (G:A)$. Thus, it suffices to show that $GL_n(FA)$ has bounded exponent. Taking any matrix $M \in GL_n(FA)$, we let H be the subgroup generated by the supports of the entries of M and M^{-1}. Then H is a finite abelian p'-group, hence FH is a direct sum of fields. In fact, if m is the exponent of A, then each such field will be of the form $F(\xi^r)$, where ξ is a primitive mth root of unity and r is a positive integer. Letting $k = |GL_n(F(\xi))|$, we see that $M^k = 1$. As k depends only upon the exponent of A, we now know that $\mathscr{U}(FG)$ satisfies $x^k = 1$, as required. $\qquad \square$

1.4 Semiprime Group Rings

Let us now discuss the result of Giambruno et al. [40] for groups with elements of infinite order. We will consider semiprime group rings in this section, and the general case in the next section.

It is worth noting that for any such result, a restriction will be required for the sufficiency, pending a positive answer to the following conjecture due to Kaplansky (see [53]).

Conjecture 1.4.1. If G is a torsion-free group and F a field, then the only units in FG are trivial; that is, λg, with $0 \neq \lambda \in F$ and $g \in G$.

This is one of the most famous (and stubborn) problems in group rings. We will see that if $\mathscr{U}(FG)$ satisfies a group identity, then the torsion elements of G form a subgroup, T. For the converse, we must find a suitable restriction upon G/T.

Recall that a group G is said to be a u.p. (unique product) group if, for every pair of nonempty finite subsets S_1 and S_2 of G, there exists an element $g \in G$ that can be uniquely written as $g = s_1 s_2$, with each $s_i \in S_i$. Furthermore, G is said to be a t.u.p. (two unique product) group if, in addition, whenever either S_1 or S_2 has more than one element, then there must be at least two such elements g. In fact, these two concepts are equivalent, as Strojnowski proved in [102].

Proposition 1.4.2. *Every u.p. group is a t.u.p. group.*

Proof. Let G be a u.p. group, and suppose that S_1 and S_2 are nonempty finite subsets, at least one of which has more than one element, such that only one element, $s_1 s_2$, of $S_1 S_2$ can be expressed uniquely in this form. Replacing S_1 with $s_1^{-1} S_1$ and S_2 with $S_2 s_2^{-1}$, we may assume that $1 \in S_1 \cap S_2$, and 1 is the only element of $S_1 S_2$ that can be expressed in only one way as the product of an element of S_1 with one of S_2.

Let $A_1 = S_2^{-1} S_1$ and $A_2 = S_2 S_1^{-1}$. Then A_1 and A_2 are nonempty finite sets. We claim that no element of $A_1 A_2$ can be written uniquely as a product of an element of A_1 with an element of A_2. This contradicts the fact that G is a u.p. group and completes the proof.

Take $a_i \in A_i$. Then $a_1 = t_2^{-1} t_1$ and $a_2 = u_2 u_1^{-1}$, with $t_i, u_i \in S_i$. Suppose that $t_1 \neq 1$ or $u_2 \neq 1$. Then the expression $t_1 u_2$ is not unique, so say $t_1 u_2 = t_1' u_2'$ gives another expression as a product of an element of S_1 with an element of S_2. But then $a_1 a_2 = (t_2^{-1} t_1')(u_2' u_1^{-1})$ shows that the expression $a_1 a_2$ is not unique.

Thus, let $t_1 = u_2 = 1$. Suppose that $t_2 \neq 1$ or $u_1 \neq 1$. Then the expression $u_1 t_2$ is not unique, so say $u_1 t_2 = u_1' t_2'$ is another expression as a product of an element of S_1 with an element of S_2. But then $a_1 a_2 = t_2^{-1} u_1^{-1} = (u_1 t_2)^{-1} = ((t_2')^{-1} 1)(1(u_1')^{-1})$ shows that the expression $a_1 a_2$ is not unique.

Finally, if $t_1 = t_2 = u_1 = u_2 = 1$, then we simply note that either S_1 or S_2 contains more than just the identity. Without loss of generality, say $1 \neq s \in S_1$. Then $1 \cdot 1 = a_1 a_2 = (1s)(1s^{-1})$, so the expression $1 \cdot 1$ is not unique. We are done. □

We recall that a group G is said to be right ordered if there is a linear order $<$ on the elements of G such that $a < b$ implies $ac < bc$ for all $a, b, c \in G$. If, in addition, $a < b$ implies $ca < cb$, then G is said to be ordered. The following observation is straightforward.

Lemma 1.4.3. *Every right ordered group G is a u.p. group.*

Proof. Let S_1 and S_2 be nonempty finite subsets of G, and let a be the largest element of S_1. Then let ab be the largest of the elements of aS_2. If $s_1 \in S_1$, $s_2 \in S_2$, then since $s_1 \leq a$, we have

$$s_1 s_2 \leq a s_2 \leq ab,$$

by choice of b. Furthermore, if either $s_1 \neq a$ or $s_2 \neq b$, then at least one of these inequalities is strict. That is, ab has a unique representation as an element of $S_1 S_2$. □

Now, free groups and torsion-free nilpotent groups are known to be ordered (see Corollary 13.2.8 and Lemma 13.1.6 in [82]). Furthermore, if G has a series of subgroups $1 = G_0 \leq G_1 \leq \cdots \leq G_m = G$, with each G_i normal in G_{i+1} and G_{i+1}/G_i

torsion-free abelian, then G is right ordered (see [82, Lemma 13.1.6]). Thus, these are examples of u.p. groups.

Also, if G is a t.u.p. group and F is any field, it is easy to see that the units of FG are trivial. Indeed, if $\alpha, \beta \in \mathscr{U}(FG)$, then let S_1 and S_2 be the supports of α and β, respectively. If either set contains more than one element, then we will get at least two group elements each appearing precisely once when we calculate $\alpha\beta$, so $\alpha\beta \neq 1$. In the same way, we can see that FG cannot have any zero divisors.

In order to begin the proof of the semiprime case, we need a couple of results on FC-groups. First, we have the well-known Dietzmann's lemma.

Lemma 1.4.4. *Let G be any group. Then every finite set of torsion elements in $\phi(G)$ is contained in a finite normal subgroup of G. In particular, the torsion elements of $\phi(G)$ form a (normal) subgroup of G.*

Proof. Take any torsion elements $h_1, \ldots, h_m \in \phi(G)$. Then let $g_1, \ldots, g_n$ be the conjugates of the h_i, including the h_i themselves. As the subgroup generated by the g_i is normal, it is sufficient to show that this subgroup is finite.

Take any $g \in \langle g_1, \ldots, g_n \rangle$. Write $g = g_{i_1}^{k_1} \cdots g_{i_r}^{k_r}$, with each $i_j \in \{1, \ldots, n\}$ and each $k_j \in \mathbb{Z}$. While this expression is surely not unique, let us choose such an expression with r minimal. We define a lexicographic ordering upon the r-tuples $(i_1, \ldots, i_r)$ as follows: namely, $(i_1, \ldots, i_r) < (i_1', \ldots, i_r')$ if there exists an s, $1 \leq s \leq r$, such that $i_j = i_j'$ whenever $j < s$, and $i_s < i_s'$. Choose our expression for g in such a way that the sequence $(i_1, \ldots, i_r)$ is minimal. Write $y_j = g_{i_j}^{k_j}$.

We claim that the i_j are all distinct. If not, then suppose that $i_c = i_d$, with $c < d$. But then

$$g = y_1 \cdots y_r = y_1 \cdots y_{c-1} y_c y_d y_{c+1}^{y_d} \cdots y_{d-1}^{y_d} y_{d+1} \cdots y_r.$$

However, $y_j^{y_d} = (g_{i_j}^{y_d})^{k_j}$ for each j. Furthermore, $g_{i_j}^{y_d} \in \{g_1, \ldots, g_n\}$. Thus, we again have an expression of length r. But $y_c y_d = g_{i_c}^{k_c + k_d}$. Thus, these two terms collapse into one, and we obtain a shorter expression for g, contradicting the minimality of r and proving the claim.

We now claim that $i_1 < i_2 < \cdots < i_r$. If not, then let us say that $i_j > i_{j+1}$. But in this case,

$$g = y_1 \cdots y_r = y_1 \cdots y_{j-1} y_{j+1} y_j^{y_{j+1}} y_{j+2} \cdots y_r.$$

Once again, we have an expression of length r, but since $i_{j+1} < i_j$, this new expression is lexicographically smaller than the old one. This gives a contradiction and proves the claim.

Therefore, every element of $\langle g_1, \ldots, g_n \rangle$ can be written in the form $g_1^{l_1} \cdots g_n^{l_n}$ for various integers l_i. That is, the order of $\langle g_1, \ldots, g_n \rangle$ is at most $o(g_1) \cdots o(g_n)$. The fact that the torsion elements of $\phi(G)$ form a normal subgroup now follows immediately. $\square$

We also need the following group-theoretic result, which is [89, 14.5.10].

Proposition 1.4.5. *Every FC-group is isomorphic to a subgroup of $H \times K$, where H is torsion-free abelian and K is locally finite.*

Let us now begin the semiprime case. When the group contains an element of infinite order, the finite field case can often be circumvented by means of the following lemma.

Lemma 1.4.6. *Let G be a group containing an element of infinite order, and F a field. Suppose that FG satisfies a nondegenerate multilinear GPI. Then there exists $\alpha \in FG$ so that $F[\alpha]$ is an infinite central subring of FG containing no zero divisors in FG.*

Proof. By Proposition 1.2.15, $(G : \phi(G)) < \infty$. As in the preceding proposition, we regard $\phi(G)$ as a subgroup of $H \times K$. Since G has an element of infinite order, so must a subgroup of finite index, so let $g \in \phi(G)$ have infinite order. Consider the (finite) collection of conjugates of g in G. As all of them lie in $\phi(G)$, and K is locally finite, we may choose a positive integer n so that $h^n \in H$ for every conjugate h of g (including g). Then replacing g with g^n, we may assume that all conjugates of g lie in H.

Let $\alpha \in F(H \cap \phi(G))$ be the sum of the conjugates of g in G. Then α is central in FG. Since H is torsion-free abelian, it is an ordered group, so we can see that the positive powers of α are all distinct. Indeed, we simply observe that for $m \geq 0$, the largest element appearing in the support of α^m is h^m, where h is the largest conjugate of g. Thus, $F[\alpha]$ is an infinite central subring of FG.

Now, since $H \cap \phi(G)$ is torsion-free abelian, and hence a u.p. group, we know that $F(H \cap \phi(G))$ has no zero divisors. Let X be a right transversal of $H \cap \phi(G)$ in G. Then if $\beta \in F(H \cap \phi(G))$ and $\gamma \in FG$, let us write $\gamma = \gamma_1 g_1 + \cdots + \gamma_k g_k$, with $\gamma_i \in F(H \cap \phi(G))$ and the g_i distinct elements of X. We see that if $\beta\gamma = 0$, then each $\beta\gamma_i = 0$. Thus, if $\beta \neq 0$, then each $\gamma_i = 0$. It follows that $F[\alpha]$ contains no zero divisors in FG, as required.   □

Thus, under the conditions described in the lemma, we may regard FG as an $F[\alpha]$-algebra.

We will need the following ring-theoretic result due to Herstein and Levitzki.

Proposition 1.4.7. *Let R be any ring. If R has a nonzero left (or right) ideal that is nil of bounded exponent, then R has a nonzero nilpotent ideal.*

Proof. Let I be a nil left ideal of exponent $n \geq 2$. (The proof for right ideals is the same.) Choose $a \in I$ with $a^{n-1} \neq 0$. Fix any $r \in R$, and let $s = ra^{n-1}$. Then $sa = ra^n = 0$. Also, $s \in I$, hence $s^n = 0$. Now, since $s + a \in I$, we have $(s+a)^n = 0$. Expanding this expression and dropping any term involving sa, as well as the terms s^n and a^n, we have $\sum_{i=1}^{n-1} a^{n-i} s^i = 0$. But notice that for all $i \in \{2, 3, \ldots, n-1\}$, we have $a^{n-i} s^i = (a^{n-i} s^{i-2} r) a^{n-1} s \in a R a^{n-1} s$. Thus,

$$0 = \sum_{i=1}^{n-1} a^{n-i} s^i = a^{n-1} s + ar' a^{n-1} s = (1 + ar') a^{n-1} s$$

for some $r' \in R$. But $(ar')^{n+1} = a(r'a)^n r' = 0$, since $r'a \in I$. Thus, ar' is nilpotent, and $1 + ar'$ is a unit. Therefore, $0 = a^{n-1} s = a^{n-1} r a^{n-1}$ for all $r \in R$. It follows immediately that $(Ra^{n-1}R)^2 = 0$. We are done.   □

The following lemma is from [31].

Lemma 1.4.8. *Let R be a semiprime D-algebra, where D is an infinite commutative F-algebra having no zero divisors in R. Suppose that $\mathscr{U}(R)$ satisfies a group identity. If $a, b, c \in R$, $bc = 0$ and a is nilpotent, then $bac = 0$.*

Proof. By Lemma 1.2.11, it suffices to assume that $a^2 = 0$. Let $f(x) \in F[x]$ be the polynomial from Lemma 1.2.3. For any $r \in R$, we have $(crb)^2 = 0$, hence $f(acrb) = 0$. In particular, if $\lambda \in D$, then $0 = f(ac\lambda rb) = f(\lambda acrb)$. Since there are infinitely many such λ, Lemma 1.2.4 tells us that $(acrb)^n = 0$, where n is the degree of f. Thus, $(bacr)^{n+1} = 0$ for all $r \in R$. That is, $bacR$ is a nil right ideal of bounded exponent. If $bacR \neq 0$, then by the previous proposition, R contains a nonzero nilpotent ideal, contradicting semiprimeness. Thus, $bac = 0$.   □

Let us tackle the semiprime case in general. Suppose that F is a field, G is a group with an element of infinite order, FG is semiprime, and $\mathscr{U}(FG)$ satisfies a group identity. Take $\alpha, \beta, \gamma \in FG$ such that α is nilpotent and $\beta\gamma = 0$. We claim that $\beta\alpha\gamma = 0$. By Lemma 1.2.11, we may assume that $\alpha^2 = 0$. Consider the polynomial $f(x)$ from Lemma 1.2.3. Now, $(\gamma\rho\beta)^2 = 0$ for all $\rho \in FG$, hence $f(\alpha\gamma\rho\beta) = 0$. Therefore, $\beta f(\alpha\gamma\rho\beta)\alpha\gamma\rho = 0$, and this is a nonzero polynomial in $\beta\alpha\gamma\rho$. That is, $I = \beta\alpha\gamma FG$ satisfies a nonzero polynomial of degree n. If $I^n \neq 0$, then by Lemma 1.2.16, FG satisfies a nondegenerate multilinear GPI, and then Lemmas 1.4.6 and 1.4.8 establish the claim. If $I^n = 0$, then since FG is semiprime, Proposition 1.4.7 tells us that $I = 0$, and therefore $\beta\alpha\gamma = 0$.

By Lemma 1.2.10, every idempotent in FG is central. As usual, we write T for the set of torsion elements of G. Let char $F = p$, and write P for the set of p-elements in G, and Q for the set of p'-elements. (If $p = 0$, then let $P = 1$ and $Q = T$.) By Lemma 1.2.12, we see that $T = Q \times P$ and Q is abelian. In particular, T is a (normal) subgroup of G, and since $\mathscr{U}(FG)$ satisfies a group identity, so does G and, therefore, G/T. Furthermore, every subgroup of P is normal in T.

We recall that a nonabelian group in which every subgroup is normal is called a Hamiltonian group. By the Dedekind–Baer theorem (see [89, 5.3.7]), the Hamiltonian groups are precisely those of the form $Q_8 \times E \times O$, where

$$Q_8 = \langle x, y | x^4 = 1, x^2 = y^2, x^{-1}yx = y^{-1} \rangle$$

is the quaternion group of order 8, E is an elementary abelian 2-group, and O is an abelian group in which every element has odd order.

We claim that $P = 1$. We know that P is abelian or Hamiltonian. If it is Hamiltonian, then as it is a p-group, we must have $p = 2$, and $P \simeq Q_8 \times E$, where E is an elementary abelian 2-group. Thus, in any case, the elements of order p are central in P, hence in T.

Take $g \in P$ of order p and let h be an element of infinite order. We claim that the set $\{g^{h^i} : i \geq 0\}$ is finite. If $g^h \in \langle g \rangle$, then $g^{h^i} \in \langle g \rangle$ for all $i \geq 0$, so the claim is proved. As the elements of order p are central in T, we may therefore assume that we have a direct product $\langle g \rangle \times \langle g^h \rangle$. Similarly, if $g^{h^2} \in \langle g \rangle \times \langle g^h \rangle$, then we would have

$g^{h^i} \in \langle g \rangle \times \langle g^h \rangle$ for all $i \geq 0$, and the claim would be proved. Thus, we have a direct product $\langle g \rangle \times \langle g^h \rangle \times \langle g^{h^2} \rangle$, and in the same way we have $\langle g \rangle \times \langle g^h \rangle \times \langle g^{h^2} \rangle \times \langle g^{h^3} \rangle$.

Notice that $(\hat{g}h^{-1}(g-1))^2 = \widehat{g^{h^2}}(g^{h^2} - 1) = 0$. Thus, we have

$$\widehat{g^{h^2}}\hat{g}h^{-1}(g-1)(g^{h^2} - 1) = 0.$$

That is,

$$\widehat{g^{h^2}}\hat{g}(g^h - 1)(g^{h^3} - 1)h^{-1} = 0.$$

Therefore,

$$\widehat{g^{h^2}}\hat{g}(g^h - 1)(g^{h^3} - 1) = 0.$$

But the terms in this last product lie in $F\langle g^{h^2} \rangle$, $F\langle g \rangle$, $F\langle g^h \rangle$ and $F\langle g^{h^3} \rangle$, respectively. As this product is direct, one of the terms must be zero, and we have a contradiction. That is, g has only finitely many distinct conjugates by positive powers of h, and similarly for each power of g. As all of these conjugates have order p, they commute, hence H, the normal closure of $\langle g \rangle$ in $\langle g, h \rangle$, is a finite p-group. Thus, by Lemma 1.1.1, $\Delta(\langle g, h \rangle, H)$ is nilpotent. In particular, $(g-1)h$ is nilpotent for all elements g of order p and all elements h of infinite order.

Since FG is semiprime, $\langle g \rangle$ cannot be normal in G, by Proposition 1.2.9. In particular, there exists an element g' with order p that lies outside of $\langle g \rangle$. Now, $(g'-1)\hat{g'} = 0$, and $(g-1)h^{-1}$ is nilpotent. Thus, $(g'-1)(g-1)h^{-1}\hat{g'} = 0$. That is,

$$(g'-1)(g-1)\widehat{(g')^h}h^{-1} = 0,$$

hence

$$(g'-1)(g-1)\widehat{(g')^h} = 0.$$

As we are working in some abelian subgroup K, Lemma 1.3.8 tells us that

$$(g'-1)(g-1) \in \Delta(K, \langle (g')^h \rangle).$$

Thus, working in $F\bar{K} = F(K/\langle (g')^h \rangle)$, we have $(\bar{g'}-1)(\bar{g}-1) = 0$. Expanding this, we get four group elements in $\bar{K}$ on the left side. Thus, we have three cases to consider. If $\bar{g} = 1$, then $g \in \langle (g')^h \rangle$. As $\langle g \rangle$ and $\langle (g')^h \rangle$ each have order p, we get $\langle g \rangle = \langle (g')^h \rangle$. If $\bar{g} = \bar{g'}$, then $g^{-1}g' \in \langle (g')^h \rangle$, and again, we must have $\langle g^{-1}g' \rangle = \langle (g')^h \rangle$. Finally, if $\bar{g} = \overline{gg'}$, then we get $\langle g' \rangle = \langle (g')^h \rangle$. In any case, $(g')^h \in \langle g, g' \rangle$. By symmetry, $g^h \in \langle g, g' \rangle$ as well. That is, $\langle g, g' \rangle$ is normalized by all elements of infinite order. It is also central in T. Thus, we have a finite normal p-subgroup, contradicting the semiprimeness of FG.

We have established the conditions for $\mathscr{U}(FG)$ to satisfy a group identity, when FG is semiprime. The first part of the main result of [40] is

Theorem 1.4.9. *Let F be a field of characteristic $p \geq 0$ and G a group containing an element of infinite order. Let FG be semiprime. If $\mathscr{U}(FG)$ satisfies a group identity, then*

1. *the torsion elements form an abelian (normal) subgroup T of G;*
2. *if $p > 0$, then G is a p'-group;*
3. *every idempotent in FG is central; and*
4. *G/T satisfies a group identity.*

Conversely, if G satisfies the four conditions above, and G/T is a u.p. group, then $\mathscr{U}(FG)$ satisfies a group identity.

Proof. We have only to check the converse. Let X be a transversal for T in G. Choose any $\alpha, \beta \in \mathscr{U}(FG)$ and $\gamma \in \mathscr{U}(FT)$. Write $\alpha = \sum_i \alpha_i g_i$, with $\alpha_i \in FT$, $g_i \in X$. Similarly, write $\alpha^{-1} = \sum_i \alpha_i' g_i$, $\beta = \sum_i \beta_i g_i$, and $\beta^{-1} = \sum_i \beta_i' g_i$. For all g_i and g_j appearing in the above expressions, write $g_i g_j = t_{ij} h_{ij}$, with $t_{ij} \in T$, $h_{ij} \in X$. Let E be the group generated by the supports of all of the $\alpha_i, \alpha_i', \beta_i, \beta_i', \gamma$, and γ^{-1}, as well as all of the t_{ij}. Then E is a finite abelian group, and char F does not divide $|E|$.

Thus, the Wedderburn decomposition of FE is a direct sum of fields. Let e be any of the primitive idempotents of FE, and write

$$FE = FEe \oplus \cdots$$

and

$$FG = FGe \oplus \cdots$$

(recalling that every idempotent is central). Now,

$$e = (\alpha e)\left(\alpha^{-1} e\right)$$

$$= \left(\sum_i (\alpha_i e) g_i\right)\left(\sum_j (\alpha_j' e) g_j\right)$$

$$= \sum_i \sum_j \alpha_i (\alpha_j')^{g_i^{-1}} e g_i g_j$$

$$= \sum_i \sum_j \left(\alpha_i (\alpha_j')^{g_i^{-1}} t_{ij} e\right) h_{ij}.$$

If we have more than one nonzero $\alpha_i e$ or more than one nonzero $\alpha_j' e$, then since G/T is a u.p. group (hence a t.u.p. group), it follows that we have at least two h_{ij} that appear exactly once in this expression. As FEe is a field, products of nonzero coefficients are nonzero. In this case, we cannot possibly get the identity of FGe, and we have a contradiction. Thus, we know that $\alpha e = \lambda g$ for some $\lambda \in \mathscr{U}(FEe)$ and $g \in X$. Similarly, $\beta e = \mu h$ for some $\mu \in \mathscr{U}(FEe)$ and $h \in X$.

Notice also that

$$\alpha^{-1} \gamma \alpha e = g^{-1} \lambda^{-1} (\gamma e) \lambda g$$

$$= g^{-1} (\gamma e) g,$$

since FE is commutative. Since every idempotent is central, we have that $\frac{1}{|E|}\hat{E}$ is central, and therefore E is a normal subgroup. Thus, $g^{-1}(\gamma e)g \in FEe$, and therefore

$\alpha^{-1}\gamma\alpha \in \mathscr{U}(FT)$ for all $\alpha \in \mathscr{U}(FG)$ and $\gamma \in \mathscr{U}(FT)$. That is, $\mathscr{U}(FT)$ is normal in $\mathscr{U}(FG)$.

Let G/T satisfy the group identity $w(x,y) = 1$. Then we see that $w(\alpha,\beta)e = w(\lambda g,\mu h)$. Now, $\mathscr{U}(FTe)$ is a normal subgroup of $\mathscr{U}(FGe)$, hence

$$w(\alpha,\beta)e \equiv w(g,h)e \pmod{\mathscr{U}(FTe)}.$$

But $w(g,h) \in T$, so $w(\alpha,\beta)e \in \mathscr{U}(FTe)$. As this holds for all of the primitive idempotents e, we have $w(\alpha,\beta) \in \mathscr{U}(FT)$ for all $\alpha,\beta \in \mathscr{U}(FG)$. However, FT is commutative, and it follows that $\mathscr{U}(FG)$ satisfies the group identity $(w(x_1,x_2),w(x_3,x_4)) = 1$. We are done. □

From the proof above we can extract the following for later use.

Remark 1.4.10. Let F be a field and G a group such that the torsion elements form a subgroup, T, and every idempotent of FT is central in FG. Further suppose that G/T is a u.p. group and T contains no elements of order divisible by char F. Let X be a transversal to T in G. Take any $\alpha \in \mathscr{U}(FG)$ and write $\alpha = \sum_i \alpha_i g_i$, and $\alpha^{-1} = \sum_i \alpha_i' g_i$, with $\alpha_i,\alpha_i' \in FT$, $g_i \in X$. For all g_i and g_j appearing in these expressions, write $g_i g_j = th$, with $t \in T$ and $h \in X$. Let E be any finite subgroup of T containing the supports of the α_i and α_i' and all of the t, and let e be a primitive idempotent of FE. Then $\alpha e = \lambda g$ for some $\lambda \in \mathscr{U}(FEe)$ and some $g \in X$. The only comment required on the proof above is that FEe need not necessarily be a field. As FE is semisimple, FEe is a matrix ring over a division ring, but a 2×2 or larger matrix ring has noncentral idempotents. Thus, FEe is a division ring, and that is sufficient for the proof.

1.5 The General Case

The characteristic zero case having been dealt with in Theorem 1.4.9, *we will assume throughout this section that F is a field of characteristic $p \geq 2$ and G is a group containing elements of infinite order.* We write T for the set of torsion elements of G and P for the set of p-elements.

If P is finite, then all we will need to do is show that it is a group. In that case, Lemma 1.2.18 tells us that if $\mathscr{U}(FG)$ satisfies a group identity, then so does $\mathscr{U}(F(G/P))$. But $F(G/P)$ is semiprime, so the conditions of Theorem 1.4.9 apply. Conversely, if $\mathscr{U}(F(G/P))$ satisfies a group identity, then we see from Lemmas 1.1.1 and 1.2.17 that $\mathscr{U}(FG)$ satisfies a group identity as well. Let us begin with

Lemma 1.5.1. *Suppose G is not torsion, and let H be a finite subgroup of $\phi(G)$. If $\mathscr{U}(FG)$ satisfies a group identity, then H' is a p-group.*

Proof. We may as well assume that F is a finite field. The centralizer of each element of H has finite index in G. Thus, by taking a suitable power of our element of infinite order, we obtain an element $g \in G$ of infinite order that centralizes H. Thus, $H \times \langle g \rangle$ is a direct product. Furthermore, $F(H \times \langle g \rangle) \cong FH \otimes_F F\langle g \rangle$, by Lemma 1.3.19. Since H is finite, we know from Proposition 1.3.3 that $J(FH)$ is nilpotent, hence $J(FH) \otimes_F F\langle g \rangle$ is nilpotent. Thus, by Lemma 1.2.17,

$$\mathscr{U}\left((FH \otimes_F F\langle g \rangle)/(J(FH) \otimes_F F\langle g \rangle)\right)$$

satisfies a group identity. But this is $\mathscr{U}\left((FH/J(FH)) \otimes_F F\langle g \rangle\right)$. However, by Proposition 1.3.3, $FH/J(FH)$ is a direct sum of matrix rings over fields (since they are finite division rings), say $\bigoplus_i M_{n_i}(F_i)$. Therefore,

$$(FH/J(FH)) \otimes_F F\langle g \rangle \cong \bigoplus_i M_{n_i}(F_i) \otimes_F F\langle g \rangle \cong \bigoplus_i M_{n_i}(F_i\langle g \rangle).$$

Thus, each $GL_{n_i}(F_i\langle g \rangle)$ must satisfy a group identity. Since $\langle g \rangle$ is infinite cyclic, Proposition 1.2.9 tells us that $F_i\langle g \rangle$ is semiprime. But matrix rings over semiprime rings are semiprime (see [62, Proposition 10.20]). Also, since $\langle g \rangle$ is an ordered group, we know that $F_i\langle g \rangle$ has no zero divisors and so the centre of $M_{n_i}(F_i\langle g \rangle)$, with which we identify it, has no zero divisors in $M_{n_i}(F_i\langle g \rangle)$. Thus, by Lemmas 1.4.8 and 1.2.10, every idempotent in the matrix ring is central, hence each $n_i = 1$. That is, $FH/J(FH)$ is a direct sum of fields, and therefore commutative. But then $H' \le 1 + J(FH)$. As $J(FH)$ is nilpotent, it follows that every element of H' is a p-element. $\qquad\square$

We can now dispense with the case in which the nilpotent radical, $N(FG)$, is a nilpotent ideal.

Proposition 1.5.2. *Let G be a group containing an element of infinite order, and suppose that $N(FG)$ is nilpotent. Then $\mathscr{U}(FG)$ satisfies a group identity if and only if P is a finite subgroup of G and $\mathscr{U}(F(G/P))$ satisfies a group identity.*

Proof. Suppose that $\mathscr{U}(FG)$ satisfies a group identity. Let $H = \phi_p(G)$. By Proposition 1.2.22, H is finite. But then by the previous lemma, H' is a p-group. Since H/H' is abelian, we conclude that the p-elements in H/H', and hence in H, form a group. That is, H is a finite p-group. Furthermore, $\phi_p(G/H) = 1$, hence $F(G/H)$ is semiprime, by Proposition 1.2.9. By Lemma 1.2.18, $\mathscr{U}(F(G/H))$ satisfies a group identity. Thus, by Theorem 1.4.9, G/H has no p-elements. That is, $P = \phi_p(G)$.

As we observed above, the converse is easy. $\qquad\square$

Let R be a ring. We recall that R is said to be prime if, for all nonzero ideals I_1 and I_2, we have $I_1 I_2 \ne 0$. An ideal I of R is said to be prime if R/I is a prime ring. Also, the lower nilradical (or prime radical), denoted $L(R)$, is the intersection of all prime ideals in R. By [82, Lemmas 8.4.2 and 8.4.3], $L(R)$ is locally nilpotent. That is, every finitely generated subring (without identity) contained in $L(R)$ is nilpotent. In addition, by [82, Lemma 8.4.2], $R/L(R)$ is semiprime. There is another description of the elements of $L(R)$ due to Levitzki. Take any $r \in R$. Let $r_1, r_2, \ldots$ be a sequence of elements of R. Letting $s_0 = r$ and $s_{n+1} = s_n r_{n+1} s_n$ for all $n \ge 0$, we say that r is strongly nilpotent if, for any such sequence, there exists an n such that $s_n = 0$. We will need the following.

Proposition 1.5.3. *Let R be any ring. Then $L(R)$ is the set of strongly nilpotent elements of R.*

Proof. See [63, Proposition 3.2.1]. □

Also, if R is an algebra over an integral domain D, we say that R is nil generated if R is generated, as a unital D-algebra, by nilpotent elements.

Lemma 1.5.4. *Suppose that R is a D-algebra, where D is an infinite commutative F-algebra having no zero divisors in R. If R is nil generated over D and $\mathcal{U}(R)$ satisfies a group identity, then $L(R)$ is the set of all nilpotent elements in R.*

Proof. Let $\bar{R} = R/L(R)$. Since $L(R)$ is nil, we know from Lemma 1.2.17 that $\mathcal{U}(\bar{R})$ satisfies a group identity. Also, $\bar{R}$ is a semiprime D-algebra. We claim that D has no zero divisors in $\bar{R}$. Indeed, suppose that $dr \in L(R)$, where $0 \neq d \in D$, $r \in R$. Choose any sequence $r_1, r_2, \ldots \in R$. Then, constructing s_i as above, starting with $s_0 = dr$, we see that $s_n = d^{2^n} s_n'$, where s_n' is the term constructed by starting with $s_0' = r$. (This follows easily since D is central in R.) Now, d is not a zero divisor, hence $s_n = 0$ if and only if $s_n' = 0$. That is, $dr \in L(R)$ if and only if $r \in L(R)$, and the claim is proved.

We see from Lemma 1.4.8 that if $\bar{a}, \bar{b}, \bar{c} \in \bar{R}$, $\bar{b}\bar{c} = 0$ and $\bar{a}$ is nilpotent, then $\overline{bac} = 0$. Let $\bar{a}_1, \ldots, \bar{a}_n$ be nilpotent, and suppose that $\bar{b}^2 = 0$. Then we get $\overline{ba_1 b} = 0$, hence $\overline{ba_1 a_2 b} = 0$, and by induction, we have $\overline{ba_1 a_2 \cdots a_n b} = 0$. Clearly $\overline{b1b} = 0$ as well. Since R is nil generated (and, therefore, so is $\bar{R}$), we have $\overline{bRb} = 0$. Thus, $(\overline{RbR})^2 = 0$. Since $\bar{R}$ is semiprime, $\bar{b} = 0$. That is, $\bar{R}$ has no nonzero square-zero elements, and therefore no nonzero nilpotent elements. We are done. □

We can now wrap up the case where P is finite.

Proposition 1.5.5. *Let F be a field of characteristic $p > 0$ and G a group containing elements of infinite order. If $\mathcal{U}(FG)$ satisfies a group identity, and $N(FG)$ is not nilpotent, then P is a (normal) subgroup of G, and $\Delta(P)$ is locally nilpotent.*

Proof. We know from Lemma 1.2.26 that FG satisfies a nondegenerate multilinear GPI. Thus, by Lemma 1.4.6, FG is a D-algebra, where D is an infinite commutative F-algebra having no zero divisors in FG. Let H be the subgroup of G generated by P. Of course, every element of FH is a field element plus an element of $\Delta(H)$. Also, notice that if g and h are p-elements, then $gh - 1 = (g-1)(h-1) + (g-1) + (h-1)$. By induction, we see that $\Delta(H)$ is generated as a ring (without identity) by the elements $\lambda(g - 1)$, with $\lambda \in F$, $g \in P$. That is, FH is nil generated, and similarly, so is R, where R is the D-subalgebra of FG generated by H.

By Lemma 1.5.4, the nilpotent elements of R form a locally nilpotent ideal, $L(R)$. In particular, since each $g - 1 \in L(R)$, where $g \in P$, we see that $\Delta(H) \subseteq L(R)$. That is, $\Delta(H)$ is locally nilpotent. Thus, if $h \in H$, then $h - 1$ is nilpotent, hence $h \in P$. It follows that $H = P$, and the p-elements form a subgroup, with $\Delta(P)$ locally nilpotent. □

As we discussed earlier, once we know that P is a finite subgroup, the work is over. We have proved another part of the main result of [40].

Theorem 1.5.6. *Let F be a field of characteristic $p > 0$, and let G be a group that is not torsion. Suppose that the set of p-elements, P, is finite. If $\mathscr{U}(FG)$ satisfies a group identity, then*

1. P is a (finite normal) subgroup of G;
2. the torsion elements of G/P form an abelian group, T/P;
3. every idempotent in $F(G/P)$ is central;
4. G/T satisfies a group identity.

Conversely, if G/T is a u.p. group, and G satisfies the four conditions above, then $\mathscr{U}(FG)$ satisfies a group identity.

Thus, we can assume from this point on that P is an infinite group and $N(FG)$ is not nilpotent.

The proof of the next lemma borrows from Billig et al. [8] and Giambruno et al. [31]. We will need to refine it for our situation, but in general we have the following.

Lemma 1.5.7. *Let R be a D-algebra, where D is an infinite commutative F-algebra having no zero divisors in R. If $\mathscr{U}(R)$ satisfies a group identity, then there is a nonzero polynomial $f(x_1, \ldots, x_k) \in F\{x_1, x_2, \ldots\}$ satisfied by all nilpotent elements of R. Furthermore, f has the following property. Let S be an F-algebra and let s_1 and s_2 be square-zero elements of S. If char $F > 2$, then f is a polynomial in two variables and $f(s_1, s_2) = \lambda (s_1 s_2)^m$ for some $0 \neq \lambda \in F$ and some $m > 0$. If char $F = 2$, then f is a polynomial in three variables and*

$$f(s_1, s_2, s_1 + s_1 s_2 + s_2 s_1 + s_2 s_1 s_2) = (s_1 s_2)^m$$

for some m.

Proof. We know from Lemma 1.1.2 that $\mathscr{U}(R)$ satisfies $w(x, y) = 1$, where $w(x, y) = x^{u_1} y^{v_1} \cdots x^{u_n} y^{v_n}$, each u_i and v_i is nonzero, $u_1 < 0$ and $v_n > 0$. Replacing x with yx^{-1} and y with $x^{-1}y$, we may assume instead that each u_i and v_i is in $\{\pm 1, \pm 2\}$, with $u_1 = v_n = 1$.

Suppose that char $F \neq 2$. As in the proof of Lemma 1.2.26, we will work in $F\{x_1, x_2\}[[z]]$ and write

$$w(1 + x_1 z, 1 + x_2 z) - 1 = \sum_{i \geq 0} f_i(x_1, x_2) z^i,$$

where each f_i is a homogeneous polynomial of degree i.

If $r_1 \in R$ satisfies $r_1^k = 0$, then of course

$$(1 + \lambda r_1)^{-1} = 1 - (\lambda r_1) + (\lambda r_1)^2 - \cdots \pm (\lambda r_1)^{k-1}$$

for any $\lambda \in D$. In particular, if $r_2 \in R$ is nilpotent as well, then

$$0 = w(1 + \lambda r_1, 1 + \lambda r_2) - 1 = \sum_{i=0}^{l} f_i(r_1, r_2)\lambda^i,$$

where l is some suitably large number. Since there are infinitely many distinct λ, we see by a Vandermonde determinant argument (as in the proof of Lemma 1.2.4), that $f_i(r_1, r_2) = 0$ for all i. Thus, the nilpotent elements of R satisfy $f_i(x_1, x_2) = 0$ for all i.

Let us now consider S. We have

$$w(1 + s_1 z, 1 + s_2 z) = (1 + s_1 z)^{u_1}(1 + s_2 z)^{v_1} \cdots (1 + s_1 z)^{u_n}(1 + s_2 z)^{v_n}$$
$$= (1 + u_1 s_1 z)(1 + v_1 s_2 z) \cdots (1 + u_n s_1 z)(1 + v_n s_2 z).$$

Notice that the coefficient of z^{2n} which is, by definition, $f_{2n}(s_1, s_2)$, is

$$u_1 \cdots u_n v_1 \cdots v_n (s_1 s_2)^n.$$

Since $u_i, v_i \in \{\pm 1, \pm 2\}$, we have our result.

Now let char $F = 2$. In $w(x, y)$, replace x with $y_1 y_3$ and y with $y_1 y_2$. Then we have the group identity

$$w'(y_1, y_2, y_3) = y_{i_1}^{c_1} \cdots y_{i_m}^{c_m},$$

with each $i_j \in \{1, 2, 3\}$, each $i_j \neq i_{j+1}$, $i_1 = 1$, $i_m = 2$, and each $c_j \in \{\pm 1\}$. Working in $F\{x_1, x_2, x_3\}[[z]]$, write

$$w'(1 + x_1 z, 1 + x_2 z, 1 + x_3 z) - 1 = \sum_{i \geq 0} f_i(x_1, x_2, x_3)z^i,$$

with each f_i a homogeneous polynomial of degree i.

Reasoning as above, we see that $f_i(r_1, r_2, r_3) = 0$ for all i and all nilpotent elements $r_1, r_2, r_3 \in R$. Let us consider S. But we have

$$w'(1 + s_1 z, 1 + s_2 z, 1 + (s_1 + s_1 s_2 + s_2 s_1 + s_2 s_1 s_2)z) = (1 + t_1 z)^{c_1} \cdots (1 + t_m z)^{c_m},$$

where

$$t_j = \begin{cases} s_1, & \text{if } i_j = 1, \\ s_2, & \text{if } i_j = 2, \\ s_1 + s_1 s_2 + s_2 s_1 + s_2 s_1 s_2, & \text{if } i_j = 3. \end{cases}$$

Noting that s_1, s_2 and $s_1 + s_1 s_2 + s_2 s_1 + s_2 s_1 s_2$ are all square-zero, and char $F = 2$, we have

$$(1 + s_1 z)^2 = (1 + s_2 z)^2 = (1 + (s_1 + s_1 s_2 + s_2 s_1 + s_2 s_1 s_2)z)^2 = 1.$$

Thus,

$$w'(1 + s_1 z, 1 + s_2 z, 1 + (s_1 + s_1 s_2 + s_2 s_1 + s_2 s_1 s_2)z) = (1 + t_1 z) \cdots (1 + t_m z).$$

In particular,

$$f_m(s_1, s_2, s_1 + s_1 s_2 + s_2 s_1 + s_2 s_1 s_2) = t_1 t_2 \cdots t_m.$$

But observing that

$$s_1(s_1 + s_1 s_2 + s_2 s_1 + s_2 s_1 s_2)s_1 = s_1 s_2 s_1 s_2 s_1,$$
$$s_1(s_1 + s_1 s_2 + s_2 s_1 + s_2 s_1 s_2)s_2 = s_1 s_2 s_1 s_2,$$
$$s_2(s_1 + s_1 s_2 + s_2 s_1 + s_2 s_1 s_2)s_1 = s_2 s_1 s_2 s_1,$$
$$s_2(s_1 + s_1 s_2 + s_2 s_1 + s_2 s_1 s_2)s_2 = s_2 s_1 s_2,$$

and recalling that $t_1 = s_1$ and $t_m = s_2$, we see immediately that

$$f_m(s_1, s_2, s_1 + s_1 s_2 + s_2 s_1 + s_2 s_1 s_2) = (s_1 s_2)^l$$

for some positive integer l. We are done.   □

As well, we need the following easy lemma.

Lemma 1.5.8. *Let G be any group and F any field. If $\alpha \in FG$ satisfies $(g-1)\alpha = 0$ for infinitely many $g \in G$, then $\alpha = 0$.*

Proof. Suppose $\alpha \neq 0$. Taking h in the support of α, it suffices to consider αh^{-1}. Thus, we can assume that 1 is in the support of α. If $g\alpha = \alpha$, then g must be in the support of α. But the support of α is finite, and we have a contradiction.   □

Proposition 1.5.9. *Suppose $N(FG)$ is not nilpotent but $\mathcal{U}(FG)$ satisfies a group identity. Then FG satisfies a polynomial identity.*

Proof. If G is torsion, then Theorem 1.2.27 does the job, so suppose that G is not torsion. By Proposition 1.5.5, P is a group. Notice that factoring out a finite normal p-subgroup will not harm the hypotheses, given Lemmas 1.1.1 and 1.2.18, or the conclusion, given Proposition 1.1.4. Thus, if P is finite, then we may factor it out. But then we see that $N(FG)$ is nilpotent, contrary to our assumption. Therefore, P is infinite. By Lemma 1.2.26, FG satisfies a nondegenerate multilinear GPI. Thus, it follows from Proposition 1.2.15 that $(G : \phi(G)) < \infty$ and $(\phi(G))'$ is finite.

By Lemmas 1.4.6 and 1.5.7, the nilpotent elements of FG satisfy a polynomial identity. In particular, $N(F\phi(G))$ satisfies a polynomial identity which, by Lemma 1.2.14, we may take to be multilinear, say $f(x_1, \ldots, x_n)$. If we can show that $F\phi(G)$ satisfies a polynomial identity, then by Proposition 1.1.4, $\phi(G)$ has a p-abelian subgroup of finite index. Hence, so does G. Thus, we may as well assume that G is an FC-group and G' is finite.

By Lemma 1.4.4, T is a locally finite group. Considering finite subgroups, then, we see from Lemma 1.5.1 that T' is a finite p-group. Thus, we can factor it out, and assume that T is abelian. But again, since G' is finite, we see that $G' \leq T$ is abelian as well. Indeed, we can quotient out $G' \cap P$ and assume that G' is a finite abelian p'-group. But then $(G, P) \leq G' \cap P = 1$. Thus, P is central.

Now, every element of $\Delta(G, P)$ can be regarded as an element of $\Delta(G, H)$, where H is a finite central p-subgroup of G. Thus, $\Delta(G, P)$ is a sum of nilpotent ideals and, therefore, contained in $N(FG)$.

Take any $\alpha_1, \ldots, \alpha_n \in FG$. We claim that $f(\alpha_1, \ldots, \alpha_n) = 0$. For any $h_1, \ldots, h_n \in P$, we have $(1 - h_i)\alpha_i \in \Delta(G, P) \subseteq N(FG)$, hence

$$0 = f((1 - h_1)\alpha_1, \ldots, (1 - h_n)\alpha_n)$$
$$= (1 - h_1) \cdots (1 - h_n) f(\alpha_1, \ldots, \alpha_n),$$

since f is multilinear and each h_i is central. Fixing $h_2, \ldots, h_n$, we have infinitely many choices for h_1. Thus, by Lemma 1.5.8,

$$(1 - h_2) \cdots (1 - h_n) f(\alpha_1, \ldots, \alpha_n) = 0$$

and eventually we get $f(\alpha_1, \ldots, \alpha_n) = 0$. □

If P has bounded exponent, then we have established the proper conditions for $\mathscr{U}(FG)$ to satisfy a group identity. The next part of the main result of [40] is

Theorem 1.5.10. *Let F be a field of characteristic $p > 0$, and let G be a group containing elements of infinite order and infinitely many p-elements. Suppose that the p-elements are of bounded exponent. If $\mathscr{U}(FG)$ satisfies a group identity, then*

1. *G has a p-abelian normal subgroup of finite index;*
2. *the p-elements of G form a (normal) subgroup P of bounded exponent;*
3. *the torsion elements of G/P form an abelian group, T/P;*
4. *every idempotent of $F(G/P)$ is central;*
5. *G/T satisfies a group identity.*

Conversely, if G satisfies the above five conditions and G/T is a u.p. group, then $\mathscr{U}(FG)$ satisfies a group identity.

Proof. Suppose that $\mathscr{U}(FG)$ satisfies a group identity. By Proposition 1.5.2, $N(FG)$ is not nilpotent. Therefore, by Proposition 1.5.9, FG satisfies a polynomial identity, so the first condition follows, and we already know that P is a subgroup. In this case, by Lemma 1.3.14, $\Delta(G, P)$ is nil of bounded exponent. Therefore, by Lemma 1.2.17, $\mathscr{U}(FG)$ satisfies a group identity if and only if $\mathscr{U}(F(G/P))$ satisfies a group identity. But $F(G/P)$ is semiprime, and Theorem 1.4.9 completes the proof. □

Thus, it remains to handle the case where P has unbounded exponent. Interestingly, this case has the nicest solution of all, and among other things, G/P turns out to be abelian. Thus, it is not necessary to impose any additional conditions upon G/T for the sufficiency.

We already know that the nilpotent elements of FG must satisfy a polynomial identity. Our next step is to get a handle on the form of the identity. We need to borrow a couple of major results from ring theory. First of all, recall that a ring R is said to be (left) primitive if it has a faithful irreducible left module. Kaplansky showed that the polynomial identities for primitive F-algebras have a particularly nice form.

Proposition 1.5.11. *Let F be a field and R a primitive F-algebra satisfying a polynomial identity. Then R is a simple algebra and satisfies precisely the same polynomial identities as $M_n(F')$, where n is some positive integer and F' is the centre of R.*

Proof. Combine Theorem 1.11.7 and Lemma 1.11.8 in [41].   □

Also recall that a ring R is said to be semiprimitive if $J(R) = 0$ or, equivalently, if R is a subdirect product of primitive rings. That is, there is a collection of primitive rings $\{R_i : i \in I\}$ such that R can be embedded in $\prod_{i \in I} R_i$, and for each $j \in I$ we have $\pi_j(R) = R_j$, where $\pi_j : \prod_{i \in I} R_i \to R_j$ is the natural projection.

We will need the celebrated Razmyslov–Kemer–Braun theorem as well.

Proposition 1.5.12. *Let F be a field and R a finitely generated F-algebra satisfying a polynomial identity. Then J(R) is nilpotent.*

Proof. See [52, Theorem 2.57].   □

We also require a couple of definitions. For any field F, let $f(x_1,\ldots,x_n)$ be a nonzero polynomial in $F\{x_1,x_2,\ldots\}$. Then the relatively free algebra of rank m, of the variety determined by f, is $R = F\{x_1,\ldots,x_m\}/I$, where I is the ideal generated by $f(a_1,\ldots,a_n)$ for all $a_i \in F\{x_1,\ldots,x_m\}$. It is clear that R satisfies f and, furthermore, every F-algebra generated by m elements that satisfies f is a homomorphic image of R.

The following proof borrows from [8], [31] and [40]. We have

Lemma 1.5.13. *Suppose that G is not torsion, $N(FG)$ is not nilpotent, and $\mathscr{U}(FG)$ satisfies a group identity. Then the nilpotent elements of FG satisfy the identity $([x_1,x_2]x_3)^{p^l} = 0$ for some l.*

Proof. From Lemmas 1.2.26 and 1.4.6 we see that FG is a D-algebra, where D is an infinite commutative F-algebra having no zero divisors in FG. Thus, Lemma 1.5.7 applies, and we let f be the polynomial identity for the nilpotent elements described therein. Let $g(x_1,x_2,x_3,x_4) = f([x_1,x_2],[x_3,x_4])$ (or

$$g(x_1,x_2,x_3,x_4) = f([x_1,x_2],[x_3,x_4],$$
$$[x_1,x_2] + [x_1,x_2][x_3,x_4] + [x_3,x_4][x_1,x_2] + [x_3,x_4][x_1,x_2][x_3,x_4])$$

if char $F = 2$). Then, let R be the relatively free algebra of rank 3 of the variety determined by g. We know that $R/J(R)$ is semiprimitive, so let us say that $R/J(R)$ embeds in $\prod_{i \in I} R_i$, with each R_i a primitive F-algebra. As $R/J(R)$ satisfies g, and $R/J(R)$ maps surjectively onto each R_i, we see that each R_i satisfies g. Fix an i.

By Proposition 1.5.11, R_i satisfies the same polynomial identities as $M_n(F')$, where F' is the centre of R_i. Suppose $n > 1$. Then if char $F > 2$, we get

$$0 = g(E_{12},E_{22},E_{21},E_{11}) = f(E_{12},E_{21}) = \lambda E_{11}^k$$

for some $k > 0$ and $\lambda \neq 0$, by Lemma 1.5.7. But this is a contradiction. Similarly, if char $F = 2$, we get

$$0 = g(E_{12}, E_{22}, E_{21}, E_{11}) = f(E_{12}, E_{21}, E_{12} + E_{12}E_{21} + E_{21}E_{12} + E_{21}E_{12}E_{21}).$$

By Lemma 1.5.7, the above equation implies once again that $(E_{12}E_{21})^k = 0$ for some k, and we have a contradiction.

Thus, $n = 1$, and $M_n(F')$ is commutative. As R_i satisfies the same identities, each R_i is commutative, and therefore $R/J(R)$ is commutative. That is, $[a,b]c \in J(R)$ for all $a, b, c \in R$. But by Proposition 1.5.12, $J(R)$ is nilpotent, hence R satisfies an identity of the form $([x_1, x_2]x_3)^{p^l} = 0$.

Let S be the D-subalgebra of FG generated by 1 and the set of nilpotent elements. Then S is nil generated, hence by Lemma 1.5.4, the nilpotent elements of S form an ideal, I, and S/I is therefore commutative. In particular, if $\alpha_i \in S$, $1 \leq i \leq 4$, then $[\alpha_1, \alpha_2], [\alpha_3, \alpha_4] \in I$, so $g(\alpha_1, \alpha_2, \alpha_3, \alpha_4) = 0$. (In the char $F = 2$ case, all of the entries substituted into f lie in I, since I is an ideal.) Hence, every subalgebra of S generated by three elements is a homomorphic image of R, and it must therefore satisfy $([x_1, x_2]x_3)^{p^l} = 0$ as well. Thus, $([\alpha_1, \alpha_2]\alpha_3)^{p^l} = 0$ for all nilpotent $\alpha_i \in FG$, as required.   □

We need to show that G' is a p-group of bounded exponent. First, let us prove

Lemma 1.5.14. *Suppose G is not torsion, $N(FG)$ is not nilpotent, and G has an abelian normal subgroup A of finite index. If $\mathscr{U}(FG)$ satisfies a group identity, then $\Delta(G, A \cap P)$ satisfies $([x_1, x_2]x_3)^{p^l} = 0$, for some l, and $(G, A \cap P)$ has bounded exponent.*

Proof. By Proposition 1.5.5, $\Delta(A \cap P)$ is locally nilpotent. Thus, since A is abelian, $\Delta(A, A \cap P) = FA\Delta(A \cap P)$ is locally nilpotent, too. It follows that for any positive integer n, $M_n(\Delta(A, A \cap P))$ is locally nilpotent. If we let $n = (G : A)$, then by Lemma 1.3.12, FG is embedded in $M_n(FA)$ and this embedding maps $\Delta(G, A \cap P)$ into $M_n(\Delta(A, A \cap P))$, since $A \cap P$ is a normal subgroup. Thus, $\Delta(G, A \cap P)$ is locally nilpotent and, in particular, nil. The previous lemma gives us the polynomial identity for $\Delta(G, A \cap P)$.

Take any $g \in G$ and $a \in A \cap P$. Notice that

$$(1 - (g,a))(1 - g^{-1}ag)g^{-1}a = g^{-1}a - g^{-1}a^{-1}gag^{-1}a - g^{-1}a^2 + g^{-1}a^{-1}gag^{-1}a^2$$
$$= g^{-1}a - ag^{-1} - g^{-1}a^2 + ag^{-1}a,$$

since elements of A commute. But this equals $[g^{-1}(a - 1), 1 - a]$. Since $\Delta(G, A \cap P)$ satisfies $([x_1, x_2]x_3)^{p^l} = 0$, we get

$$0 = ([g^{-1}(a - 1), 1 - a](a^{-1}g(a - 1)))^{p^l}$$
$$= ((1 - (g,a))(1 - g^{-1}ag)(a - 1))^{p^l}$$
$$= (1 - (g,a)^{p^l})(1 - g^{-1}a^{p^l}g)(a^{p^l} - 1).$$

Letting $b = a^{p^l}$ and $c = g^{-1}a^{p^l}g$, we note that $(g,a)^{p^l} = c^{-1}b$, since A is abelian. Thus, we have

$$0 = (1 - c^{-1}b)(1 - c)(1 - b) = 1 - b^2 + bc - bc^{-1} + b^2 c^{-1} - c.$$

That is, $1 \in \{b^2, bc, bc^{-1}, b^2 c^{-1}, c\}$. As $(G, A \cap P)$ is contained in $A \cap P$, and is therefore abelian, in order to show that it has bounded exponent, it suffices to show that $(g, a)^{p^{l+2}} = 1$. In other words, we must show that $(bc^{-1})^{p^2} = 1$.

If $c = 1$, then $b = 1$ as well, and there is nothing to do. Suppose $b^2 = 1$. Then bc^{-1} has order dividing 2. Since it is a p-element, we are done. If $bc^{-1} = 1$, there is nothing to say. Next, if $bc = 1$, then we have

$$2 - 2b^2 + b^3 - b^{-1} = 0.$$

Thus, $b^{-1} \in \{1, b^2, b^3\}$, so $o(b) = o(c) \leq 4$. Therefore, $o(bc^{-1}) \leq 4 \leq p^2$. If $b^2 c^{-1} = 1$, then we obtain the same equation, and we are done. $\qquad \square$

The next lemma is the last piece of the puzzle.

Lemma 1.5.15. *Suppose G is not torsion and $\mathscr{U}(FG)$ satisfies a group identity. If P has unbounded exponent, then G' is a p-group of bounded exponent.*

Proof. By Proposition 1.5.2, $N(FG)$ is not nilpotent. Thus, by Proposition 1.5.9, FG satisfies a polynomial identity. Hence, by Proposition 1.1.4, G has a p-abelian normal subgroup A of finite index. Since, by Lemma 1.1.1, $\Delta(G, A')$ is nilpotent, we may factor out A' and assume that A is abelian. Thus, by the previous lemma, $(G, A \cap P)$ has bounded exponent and $\Delta(G, A \cap P)$ satisfies the identity $([x_1, x_2] x_3)^{p^l} = 0$ for some l. Hence, it is sufficient to show that $(G/(G, A \cap P))'$ is a p-group of bounded exponent. We therefore factor out $(G, A \cap P)$ and assume that $A \cap P$ is central. Furthermore, P then has a central subgroup of finite index and therefore, P' is finite, by Proposition 1.3.7. Thus, we can factor out P' and assume that P is abelian.

Since P has unbounded exponent, so does $A \cap P$, being a subgroup of finite index. Thus, choose $a \in A \cap P$ of order p^{l+3}. Take any $g, h \in G$. We claim that $(g, h)^{p^{l+1}} = 1$. We have

$$0 = ([h^{-1}(a-1), g^{-1}(a-1)](gh(a-1)))^{p^l} = (a-1)^{3p^l}(1 - (g, h)^{p^l}),$$

since a is central. Multiplying by $(a-1)^{p^{l+2} - 3p^l}$, we get

$$(a^{p^{l+2}} - 1)(1 - (g, h)^{p^l}) = 0.$$

That is,

$$a^{p^{l+2}} + (g, h)^{p^l} - a^{p^{l+2}}(g, h)^{p^l} - 1 = 0.$$

If $p > 2$, then we must have $a^{p^{l+2}} = 1$ (which is not true) or $(g, h)^{p^l} = 1$, which establishes our claim. If $p = 2$, we have also to consider the possibility that $a^{2^{l+2}}(g, h)^{2^l} = 1$. But then $(g, h)^{2^{l+1}} = a^{-2^{l+3}} = 1$. Our claim is established. Since G' is a subgroup of P, which is abelian, and G' is generated by commutators of bounded exponent, it follows that G' has bounded exponent. $\qquad \square$

We close the chapter with the final part of the main result of Giambruno et al. [40].

Theorem 1.5.16. *Let F be a field of characteristic $p > 0$. Let G be a group containing elements of infinite order and let the p-elements of G have unbounded exponent. Then $\mathscr{U}(FG)$ satisfies a group identity if and only if G has a p-abelian normal subgroup A of finite index and G' is a p-group of bounded exponent.*

Proof. Suppose $\mathscr{U}(FG)$ satisfies a group identity. By Proposition 1.5.2, $N(FG)$ is not nilpotent. Thus, by Proposition 1.5.9, FG satisfies a polynomial identity. The existence of A then follows from Proposition 1.1.4. We obtain from the previous lemma that G' is a p-group of bounded exponent.

Let us verify the converse. Surely $\mathscr{U}(F(G/G'))$ satisfies $(x,y) = 1$. That is, $\mathscr{U}(FG/\Delta(G,G'))$ satisfies $(x,y) = 1$. But by Proposition 1.1.4, FG satisfies a polynomial identity and therefore, by Lemma 1.3.14, $\Delta(G,G')$ is nil of bounded exponent. Let us say that this exponent is at most p^k. Then by Lemma 1.2.17, $\mathscr{U}(FG)$ satisfies $(x,y)^{p^k} = 1$. We are done.   □

In fact, when P has unbounded exponent, it follows from the proof of the last theorem that if $\mathscr{U}(FG)$ satisfies a group identity, then it satisfies an identity of the form $(x,y)^{p^k} = 1$ for some k.

Chapter 2
Group Identities on Symmetric Units

2.1 Introduction

We now turn our attention from the unit group of the group ring to the set of symmetric units. Some definitions are in order. Let R be a ring. Then an involution is a function $* : R \to R$ satisfying $(r+s)^* = r^* + s^*$, $(rs)^* = s^*r^*$ and $(r^*)^* = r$ for all $r, s \in R$. Classic examples of involutions include conjugation on the complex numbers and the transpose function $(r_{ij}) \mapsto (r_{ji})$ on the ring of $n \times n$ matrices over any field.

The elements of R fixed by an involution are said to be symmetric with respect to that involution, and we write R^+ for the set of symmetric elements. From time to time, we will also consider the set of skew elements, $R^- = \{r \in R : r^* = -r\}$. If R is an F-algebra with char $F \neq 2$, and $*$ fixes F elementwise, then R is easily seen to be a direct sum (as vector spaces) of R^+ and R^-, since $r = (\frac{r+r^*}{2}) + (\frac{r-r^*}{2})$, for all $r \in R$. We write $\mathscr{U}^+(R)$ for the set of symmetric units. It seems natural to explore the extent to which the symmetric units determine the structure of the unit group.

If F is a field and G a group, then FG has a natural involution given by

$$\left(\sum_{g \in G} \alpha_g g \right)^* = \sum_{g \in G} \alpha_g g^{-1}.$$

This is the involution upon which we focus in this book. In Chapter 7, we will briefly examine other involutions, but until then, *whenever we mention $*$ on FG, we will assume that it is this involution*. It is easy to see that if char $F \neq 2$, then the symmetric elements are the F-linear combinations of $g + g^{-1}$, $g \in G$. When F has characteristic 2, we must consider the elements of order 1 or 2 as well, since we cannot take $\frac{g+g^{-1}}{2}$. Nearly all of the results for the symmetric units assume that char $F \neq 2$.

An obvious first question is this: If $\mathscr{U}^+(FG)$ satisfies a group identity, does it follow that $\mathscr{U}(FG)$ satisfies a group identity? The answer, in general, is no. Recall, for instance, that if char $F = 0$ and G is torsion, then $\mathscr{U}(FG)$ satisfies a group

G.T. Lee, *Group Identities on Units and Symmetric Units of Group Rings*,
Algebra and Applications 12, DOI 10.1007/978-1-84996-504-0_2,
© Springer-Verlag London Limited 2010

identity if and only if G is abelian (see Corollary 1.2.21). However, we have the following simple observation. Recall that a Hamiltonian 2-group has the form $Q_8 \times E$, where Q_8 is the quaternion group and E is an elementary abelian 2-group.

Lemma 2.1.1. *Let F be any field and G a Hamiltonian 2-group. Then $(FG)^+$ is the centre of FG.*

Proof. Take $\alpha = \sum_{g \in G} \alpha_g g \in FG$. Then α is central if and only if

$$\alpha = \sum_{g \in G} \alpha_g h^{-1} g h$$

for all $h \in G$; that is, if and only if $\alpha_g = \alpha_{gh}$ for all $g, h \in G$. But it is easy to see that every element of order 1 or 2 in G is central, and if $a \in G$ does not have order 1 or 2, then the conjugates of a are a and a^{-1}. Thus, the central elements are precisely the linear combinations of the elements of order 1 or 2, and the terms $a + a^{-1}$ for all other group elements. But these are the symmetric elements of FG.   □

Thus, in this case, the symmetric units commute, and therefore they satisfy a group identity. We will see that many of the results concerning the symmetric elements break down into two cases: groups containing the quaternions and groups not containing them.

Of course, if the symmetric units do not commute, then they do not form a group, since $(\alpha\beta)^* = \beta^*\alpha^* = \beta\alpha$, for all $\alpha, \beta \in (FG)^+$. (See Bovdi et al. [18] and Bovdi [17] for a discussion of the groups G such that $\mathscr{U}^+(FG)$ is a group.) But, we can still ask if $\mathscr{U}^+(FG)$ satisfies a group identity.

In this chapter, we will present the results of Giambruno et al. [39] (for torsion groups) and Sehgal and Valenti [96] (for groups with elements of infinite order). Their results provide a complete classification of the groups G such that $\mathscr{U}^+(FG)$ satisfies a group identity, whenever F is an infinite field of characteristic different from 2 (subject to the same condition on G modulo its torsion elements that was needed in the previous chapter). The problem is currently open for finite fields and fields of characteristic 2.

An interesting question along the way is this: Is there an analogue for Hartley's conjecture? We will need the notion of a $*$-polynomial identity. We can define an involution on the free algebra $F\{x_1, x_2, \ldots\}$ by letting $x_1^* = x_2$, $x_3^* = x_4$, and so forth. Renumbering, we obtain the free algebra with involution $F\{x_1, x_1^*, x_2, x_2^*, \ldots\}$. Let R be an F-algebra with involution; that is, we insist that $(\lambda r)^* = \lambda r^*$ for all $\lambda \in F$, $r \in R$. We say that R satisfies a $*$-polynomial identity if there is a nonzero polynomial $f(x_1, x_1^*, \ldots, x_n, x_n^*) \in F\{x_1, x_1^*, \ldots\}$ such that $f(r_1, r_1^*, \ldots, r_n, r_n^*) = 0$ for all $r_1, \ldots, r_n \in R$. As part of the Giambruno et al. result, we will see that if G is torsion and $\mathscr{U}^+(FG)$ satisfies a group identity, then $(FG)^+$ satisfies a polynomial identity. In particular, then, FG must satisfy a $*$-polynomial identity. (Indeed, if $f(x_1, \ldots, x_n)$ is a polynomial identity for $(FG)^+$, then $f(x_1 + x_1^*, \ldots, x_n + x_n^*)$ is a $*$-polynomial identity for FG.) But more can be said, due to this classical result of Amitsur.

Proposition 2.1.2. *Let F be a field and R an F-algebra with involution (with or without an identity). If R satisfies a $*$-polynomial identity, then R satisfies a polynomial identity.*

Proof. See [48, p. 195]. □

Thus, Hartley's conjecture is true for the symmetric units as well.

We can always assume that our group identity is an identity in two variables. We do have to be a little bit careful in our argument, in that we must make sure that we are only making symmetric substitutions.

Lemma 2.1.3. *Let R be a ring with involution and suppose that $\mathscr{U}^+(R)$ satisfies a group identity. Then $\mathscr{U}^+(R)$ satisfies an identity of the form $x^{i_1}y^{j_1}x^{i_2}\cdots y^{j_{m-1}}x^{i_m}=1$, where each exponent is a nonzero integer and $i_1 > 0$.*

Proof. Suppose that $\mathscr{U}^+(R)$ satisfies $w(x_1,\ldots,x_n)=1$. Notice that if $u,v\in\mathscr{U}^+(R)$, then also $u^i v u^i\in\mathscr{U}^+(R)$ for any positive integer i. Thus, we can substitute $x^i y x^i$ for x_i, and we obtain a group identity of the required form, replacing x with x^{-1} if necessary in order to ensure that $i_1 > 0$. □

It is useful to know about the involutions on matrix rings as well. Fortunately, these have been extensively studied. We have already seen that the transpose function is an involution on $M_n(F)$ for any field F. If n is even, there is another important involution, called the symplectic involution. If $A_{11},A_{12},A_{21},A_{22}\in M_{n/2}(F)$, then we let

$$\begin{pmatrix} A_{11} & A_{12} \\ A_{21} & A_{22} \end{pmatrix}^s = \begin{pmatrix} A_{22}^t & -A_{12}^t \\ -A_{21}^t & A_{11}^t \end{pmatrix},$$

where t is the usual matrix transpose. In fact, by [57, Propositions 2.19 and 2.20], all of the involutions on $M_n(F)$ can be expressed in terms of these two involutions. But up to an isomorphism respecting the involution, we can simplify that result and allow for division rings as well. The following is a classical result due to Kaplansky.

Proposition 2.1.4. *Let D be a division ring of characteristic different from 2 and n a positive integer. Let $*$ be any involution on $M_n(D)$. Then, up to an automorphism θ of $M_n(D)$ satisfying $\theta(A^*) = (\theta(A))^*$ for all $A\in M_n(D)$, we have either:*

1. there exist an involution $^-$ of D and an invertible diagonal matrix

$$U = \begin{pmatrix} u_{11} & 0 & \cdots & 0 \\ 0 & u_{22} & \cdots & 0 \\ \vdots & \vdots & \ddots & \vdots \\ 0 & 0 & \cdots & u_{nn} \end{pmatrix}$$

such that $\bar{u}_{ii} = u_{ii}$ for all i, and if $A = (a_{ij})\in M_n(D)$, then $A^* = U^{-1}BU$, where $b_{ij} = \bar{a}_{ji}$ for all i and j; or
2. D is a field, n is even, and $*$ is the symplectic involution.

Proof. See [6, Theorem 4.6.8 and Lemma 4.6.11]. □

Note that involutions of the sort described in the first part of the preceding proposition are said to be of transpose (or orthogonal) type.

We will begin with some results concerning semiprime rings with involution, and then move on to group rings of finite groups and the general torsion case. Following that, we will discuss semiprime group rings of nontorsion groups, and then the general case.

2.2 Semiprime Rings

*Throughout this section, F is an infinite field of characteristic different from 2 and R is an F-algebra having an involution * such that * fixes F elementwise.*

Let us begin with the following lemma.

Lemma 2.2.1. *Suppose $\mathscr{U}^+(R)$ satisfies the group identity $w(x,y) = 1$, as in Lemma 2.1.3. Then there is a positive integer n, depending only upon w, such that*

1. *if $a \in R$ is square-zero, then $(aa^*)^n = 0$; and*
2. *if $b,c \in R^+$ are square-zero, then $(cbcd)^n = 0$ for all $d \in R^+$.*

Proof. Suppose $a \in R$, $a^2 = 0$. Then for any $\lambda \in F$, we have $(1+\lambda a)(1+\lambda a^*) \in \mathscr{U}^+(R)$, as its inverse is easily seen to be $(1-\lambda a^*)(1-\lambda a)$. In the same way, $(1+\lambda a^*)(1+\lambda a) \in \mathscr{U}^+(R)$. Thus,

$$w((1+\lambda a)(1+\lambda a^*), (1+\lambda a^*)(1+\lambda a)) = 1.$$

Expanding this, we obtain $\sum_{i=1}^m p_i(a,a^*)\lambda^i = 0$, where each p_i is a polynomial in a and a^*. We now apply a Vandermonde argument similar to Lemma 1.2.4. Indeed, since F is infinite, let $\lambda_1,\ldots,\lambda_{m+1}$ be distinct elements of F. We have

$$\begin{pmatrix} 1 & \lambda_1 & \lambda_1^2 & \cdots & \lambda_1^m \\ 1 & \lambda_2 & \lambda_2^2 & \cdots & \lambda_2^m \\ \vdots & \vdots & \vdots & \ddots & \vdots \\ 1 & \lambda_{m+1} & \lambda_{m+1}^2 & \cdots & \lambda_{m+1}^m \end{pmatrix} \begin{pmatrix} 0 \\ p_1(a,a^*) \\ \vdots \\ p_m(a,a^*) \end{pmatrix} = \begin{pmatrix} 0 \\ 0 \\ \vdots \\ 0 \end{pmatrix}.$$

As a Vandermonde matrix is invertible, we get that $p_i(a,a^*) = 0$ for all i. However, we note that in

$$w((1+\lambda a)(1+\lambda a^*), (1+\lambda a^*)(1+\lambda a)),$$

we have at most two consecutive instances of $1+\lambda a$ or $1+\lambda a^*$ (and at most two consecutive instances of their inverses, $1-\lambda a$ and $1-\lambda a^*$). Since $(1+\lambda a)^2 = 1+2\lambda a$, and similarly for the other terms, we know that the leading coefficient is $p_m(a,a^*) = \pm 2^k(aa^*)^l a^j$, where $k \geq 0$, $l \geq 1$, and $j \in \{0,1\}$. Thus, $(aa^*)^{l+1} = 0$, and the first part is established.

For the second part, we note that for any $\lambda \in F$,

$$(1 + \lambda c)(1 + \lambda b)(1 + \lambda c), (1 + \lambda b)(1 + \lambda c)(1 + \lambda b) \in \mathcal{U}^+(R).$$

Substituting these symmetric units into w, we repeat the argument above and learn that $(cb)^r = 0$ for some positive integer r. But notice that cdc is symmetric and square-zero. Thus, replacing c with cdc, we get $(cdcb)^r = 0$, hence $(cbcd)^{r+1} = 0$. Taking n to be the larger of $l + 1$ and $r + 1$, we are done. $\qquad\square$

The next result is Theorem 1 in [39]. The proof is very nice, but it requires some machinery that is not needed elsewhere in the book. For convenience, the proof can be found in the appendix.

Proposition 2.2.2. *Let R be semiprime. Fix a central element $z \in R$ and a positive integer k. If $a \in R$ satisfies $(ab(z - a)b^*)^k = 0$ for all $b \in R$, then a is central in R.*

As an easy consequence, we have

Lemma 2.2.3. *If R is semiprime and $a \in R^+$ satisfies $(ab)^k = 0$ for all $b \in R^+$, then a is central.*

Proof. Let $b = cac^*$ for any $c \in R$. Then b is symmetric, hence $(acac^*)^k = 0$. We now apply Proposition 2.2.2, using $z = 0$. $\qquad\square$

We can now deduce

Lemma 2.2.4. *If R is semiprime and $\mathcal{U}^+(R)$ satisfies a group identity, then every symmetric idempotent of R is central.*

Proof. Let e be a symmetric idempotent. Take any $r \in R$. Then $(er(1 - e))^2 = 0$; hence, by Lemma 2.2.1, there exists a positive integer n such that

$$0 = ((er(1 - e))(er(1 - e))^*)^n = (er(1 - e)r^*e)^n.$$

It follows that $(er(1 - e)r^*)^{n+1} = 0$. Thus, applying Proposition 2.2.2 with $z = 1$, we see that e is central. $\qquad\square$

This gives us a nice consequence for semiprime group rings.

Lemma 2.2.5. *Let F be an infinite field of characteristic different from 2 and G a group, such that FG is semiprime and $\mathcal{U}^+(FG)$ satisfies a group identity. If g is an element of finite order in G, and char F does not divide $o(g)$, then $\langle g \rangle$ is a normal subgroup.*

Proof. Observe that $\frac{1}{o(g)}\hat{g}$ is a symmetric idempotent. Thus, by the previous lemma, it is central. Therefore, $\langle g \rangle$ is normal. $\qquad\square$

In particular, it follows that if char $F = 0$ and G is torsion, or if char $F = p > 2$ and G is a torsion p'-group, then G is abelian or Hamiltonian. We will refine this observation later on.

We record two other results for later use. The first is similar to Lemma 1.5.7.

Lemma 2.2.6. *Suppose that $\mathcal{U}^+(R)$ satisfies a group identity. Let S be a nil F-subalgebra of R (without identity) such that S is invariant under $*$. Then S satisfies a polynomial identity.*

Proof. Assume that the group identity is of the form $w(x,y) = 1$, as in Lemma 2.1.3. We can consider the power series ring $F\{x_1,x_2\}[[z]]$, and by Lemma 1.2.25, the elements $1 + x_1 z$ and $1 + x_2 z$ generate a free subgroup of the unit group. Thus,

$$0 \neq w(1 + x_1 z, 1 + x_2 z) - 1 = \sum_{i \geq 0} f_i(x_1, x_2) z^i,$$

and some f_m is not the zero polynomial. Choosing $s_1, s_2 \in S^+$, we know that $1 + \lambda s_i \in \mathcal{U}^+(R)$ for each i and all $\lambda \in F$. Thus,

$$0 = w(1 + \lambda s_1, 1 + \lambda s_2) - 1 = \sum_{i \geq 0} f_i(s_1, s_2) \lambda^i.$$

But, for any particular s_1 we have $(1 + s_1)^{-1} = 1 - s_1 + s_1^2 - \cdots \pm s_1^k$ for some k, and similarly for s_2. That is, there exists a $j \geq m$ such that $f_i(s_1, s_2) = 0$ for all $i > j$. Thus,

$$\sum_{i=0}^{j} f_i(s_1, s_2) \lambda^i = 0.$$

Since F is infinite, choose distinct elements $\lambda_1, \ldots, \lambda_{j+1} \in F$. Then

$$\begin{pmatrix} 1 & \lambda_1 & \lambda_1^2 & \cdots & \lambda_1^j \\ 1 & \lambda_2 & \lambda_2^2 & \cdots & \lambda_2^j \\ \vdots & \vdots & \vdots & \ddots & \vdots \\ 1 & \lambda_{j+1} & \lambda_{j+1}^2 & \cdots & \lambda_{j+1}^j \end{pmatrix} \begin{pmatrix} f_0(s_1, s_2) \\ f_1(s_1, s_2) \\ \vdots \\ f_j(s_1, s_2) \end{pmatrix} = \begin{pmatrix} 0 \\ 0 \\ \vdots \\ 0 \end{pmatrix}.$$

Since a Vandermonde matrix is invertible, each $f_i(s_1, s_2) = 0$. In particular, S^+ satisfies $f_m(x_1, x_2)$. That is, S satisfies the $*$-polynomial identity $g_m(x_1, x_1^*, x_2, x_2^*) = f_m(x_1 + x_1^*, x_2 + x_2^*)$. But then Proposition 2.1.2 tells us that S satisfies a polynomial identity.   □

Lemma 2.2.7. *Let R be semiprime and suppose that $\mathcal{U}^+(R)$ satisfies a group identity. Take any square-zero symmetric element a of R. Then:*

1. if b is a symmetric nilpotent element of R, then $aba = 0$; and
2. if $c, d \in R$ satisfy $cd = 0$, then $cad = 0$.

Proof. For the first part, let us suppose, to begin with, that $b^2 = 0$. Then by Lemma 2.2.1, $(abar)^n = 0$ for all $r \in R^+$. Thus, by Lemma 2.2.3, aba is central. But aba is also square-zero, and a semiprime ring cannot have a nonzero central nilpotent element, lest it generate a nonzero nilpotent ideal. Thus, $aba = 0$.

Let us consider the general case. If $aba \neq 0$, then let i be the largest positive integer such that $ab^i a \neq 0$. (If $b^j = 0$, then evidently $i < j$.) Take any $r \in R^+$. Then

$(b^i arab^i)^2 = b^i ar(ab^{2i}a)rab^i = 0$, and $b^i arab^i$ is symmetric. Thus, as we have just observed, $ab^i arab^i a = 0$. Hence, $(ab^i ar)^2 = 0$ for all $r \in R^+$, and by Lemma 2.2.3, $ab^i a$ is central. Once again, we cannot have nonzero central nilpotent elements, so $ab^i a = 0$, giving us a contradiction.

For the second part, suppose that s is a square-zero element of R. By Lemma 2.2.1, $(ss^*)^n = (s^*s)^n = 0$. But this implies that $(s+s^*)^{2n} = 0$. Indeed, when we expand $(s+s^*)^{2n}$, we get a number of terms in which s^2 or $(s^*)^2$ appear, and these vanish, along with $(ss^*)^n + (s^*s)^n = 0$. Thus, by the first part of the lemma, we have $a(s+s^*)a = 0$ or, equivalently, $asa = -as^*a$. Now, $(sas)^2 = 0$, so replacing s with sas, we obtain $a(sas + (sas)^*)a = 0$. Thus,

$$asasa = -as^*(as^*a) = (as^*a)sa = -asasa,$$

and, therefore, $asasa = 0$. Now, for any $r \in R$, we see that drc is square-zero, hence $adrcadrca = 0$ and $(cadr)^3 = 0$ for all $r \in R$. By Proposition 1.4.7, this cannot happen unless $cad = 0$. We are done.   □

The semiprime case for group rings of torsion groups will be resolved quickly once we have dealt with the finite group case.

2.3 Group Rings of Finite Groups

Let us suppose that G is a finite group. We once again assume that F is an infinite field. We begin with a useful general lemma.

Lemma 2.3.1. *Let R be a ring with involution, with $\frac{1}{2} \in R$, and let I be a $*$-invariant nil ideal of R. Then $\mathscr{U}^+(R/I)$ is the image of $\mathscr{U}^+(R)$ under the natural map $R \to R/I$.*

Proof. It is evident that symmetric units map to symmetric units. Take any symmetric unit $\bar{u} \in \bar{R} = R/I$. Let $\bar{v} = (\bar{u})^{-1}$. Then $uv - 1 \in I$, say $uv = 1 + r$, with $r \in I$. But r is nilpotent, so let us say that $r^k = 0$. Then we have $uv(1 - r + r^2 - \cdots \pm r^{k-1}) = 1$, hence u has a right inverse and, by the same argument, a left inverse in R. That is, $u \in \mathscr{U}(R)$. Now, $\bar{u} = \bar{u}^*$, hence $u^* = u + s$ for some $s \in I$. Thus, $u + u^* = 2u(1+t)$ for some $t \in I$. As t is nilpotent, $2u(1+t)$ is a unit, and it is clearly symmetric. Also, $\frac{1}{2}(u+u^*) = \bar{u}$. Thus, $\bar{u}$ is the image of a symmetric unit in R. We are done.   □

First of all, suppose that char $F = p > 2$. We must show that if $\mathscr{U}^+(FG)$ satisfies a group identity, then the p-elements of G form a group. Let us begin with

Lemma 2.3.2. *Let F be an infinite field of characteristic $p > 2$ and G a finite group. If $\mathscr{U}^+(FG)$ satisfies a group identity, then for every p-element g of G, $(g-1)^2 \in J(FG)$.*

Proof. Observe that $J(FG)$ is invariant under $*$. Indeed, one of the definitions of $J(FG)$ is that it consists of all $\alpha \in FG$ such that $1 - \beta\alpha\gamma \in \mathscr{U}(FG)$ for all $\beta, \gamma \in$

FG (see [62, Lemma 4.3]). Thus, if $\alpha \in J(FG)$, then $1 - \beta\alpha^*\gamma = (1 - \gamma^*\alpha\beta^*)^* \in \mathcal{U}(FG)$ for all $\beta, \gamma \in FG$; hence $\alpha^* \in J(FG)$. Let $R = FG/J(FG)$. Then R has an induced involution. Furthermore, by Proposition 1.3.3, $J(FG)$ is nilpotent. Thus, by the preceding lemma, $\mathcal{U}^+(R)$ is a homomorphic image of $\mathcal{U}^+(FG)$ and therefore satisfies a group identity. We claim that if $r \in R$ is nilpotent, then $rr^* = 0$. This will complete the proof, since if $g \in G$ is a p-element, then $g - 1$ is nilpotent, hence

$$(g-1)(g^{-1}-1) = (g-1)(g-1)^* \in J(FG),$$

and therefore $(g-1)^2 = -g(g-1)(g^{-1}-1) \in J(FG)$.

We know that R is semisimple, so let us write

$$R = Re_1 \oplus \cdots \oplus Re_m,$$

where each e_i is a primitive central idempotent, and $Re_i \cong M_{n_i}(D_i)$, with n_i a positive integer and D_i a division algebra.

Suppose, first of all, that e_i is symmetric. Then the projection $R \to Re_i$ induces an involution on $M_{n_i}(D_i)$. Furthermore, every symmetric unit $ae_i \in Re_i$ is the image of a symmetric unit $ae_i + (1 - e_i)$ in R. (If $(ae_i)(be_i) = e_i$, then also $(ae_i + (1-e_i))(be_i + (1-e_i)) = 1$.) Thus, the symmetric units of $M_{n_i}(D_i)$ satisfy a group identity. Considering Proposition 2.1.4, we may assume that this involution is either of transpose type or symplectic. Suppose that it is of transpose type. Then the matrix unit E_{11} is easily seen to be a symmetric idempotent. By Lemma 2.2.4, E_{11} is central. Thus, $n_i = 1$ and Re_i is a division algebra. On the other hand, if the involution is symplectic, then D_i is a field and $E_{11} + E_{n_i/2+1,n_i/2+1}$ is a symmetric idempotent and, therefore, central. Clearly, $n_i = 2$ and by definition of the symplectic involution on 2×2 matrices, the symmetric elements are simply scalar multiples of the identity matrix.

Suppose, on the other hand, that e_i is not symmetric. Then e_i^* is also a primitive central idempotent, say $e_i^* = e_j$. Notice that if $ae_i \in \mathcal{U}(Re_i)$, then ae_i is the image of $ae_i + a^*e_j + (1 - (e_i+e_j))$, and this is a symmetric unit. Thus, $GL_{n_i}(D_i)$ satisfies a group identity. By Proposition 1.2.2, D_i is a field, and by Lemma 1.2.6, $n_i = 1$. Thus, Re_i is a field.

In summary, each Re_i is a division algebra or a 2×2 matrix ring over a field, and in the latter case, it has an induced involution under which all of the symmetric elements are central. Now, suppose that $r \in R$ is nilpotent. Then so is each re_i. If Re_i is a division algebra, then $re_i = 0$. Otherwise, re_i is a nilpotent 2×2 matrix. Since the minimal polynomial of this matrix has degree at most 2, $(re_i)^2 = 0$. Thus, $r^2 = 0$. By Lemma 2.2.1, $(rr^*)^n = 0$, and by the same argument, if Re_i is a division algebra, then $rr^*e_i = 0$. If Re_i is a 2×2 matrix ring, then rr^*e_i is symmetric, hence central. But a central nilpotent matrix must be the zero matrix. Thus, $rr^* = 0$. $\square$

In order to show that the p-elements form a subgroup, we need to borrow a result about group representations. For our purposes, when we speak of a representation over a field F, we will mean a homomorphism $\rho : G \to GL_n(F)$, where G is a finite group, n (the degree) is a positive integer and F is an algebraically closed field.

Recall that ρ is said to be faithful if $\ker(\rho) = 1$. As usual, we write $SL_n(F)$ for the group of $n \times n$ matrices of determinant one over a field F, and $PSL_n(F)$ for the group obtained by factoring out the centre of $SL_n(F)$.

Proposition 2.3.3. *Let F be the algebraic closure of $\mathbb{Z}_p$, the field of p elements, where p is any prime. Let G be a finite group generated by two p-elements g and h. Suppose that G has a faithful irreducible representation ρ over F such that the minimal polynomials of $\rho(g)$ and $\rho(h)$ are quadratic. Then G has a subgroup isomorphic to $SL_2(\mathbb{Z}_p)$.*

Proof. See [45, Theorem 3.8.1]. □

Now we have

Lemma 2.3.4. *Let F be an infinite field of characteristic $p > 2$ and let G be a finite group. If, for every p-element g of G, we have $(g-1)^2 \in J(FG)$, then the p-elements of G form a (normal) subgroup.*

Proof. First of all, we can assume that F is the field mentioned in Proposition 2.3.3. Indeed, by Proposition 1.3.3, we can shrink the field to $\mathbb{Z}_p$ and then expand it to its algebraic closure. Thus, we will assume that this is indeed our field.

Next, we note that our hypothesis upon G is inherited by its subgroups and homomorphic images. Indeed, if $H \le G$, then for any p-element g of H, we have $(g-1)^2 \in J(FG) \cap FH \subseteq J(FH)$, by Proposition 1.3.3. If H is normal, then for any p-element $\bar{g}$ of $\bar{G} = G/H$, let us choose an integer k, relatively prime to p, such that g^k is a p-element. Then $(g^k - 1)^2 \in J(FG)$. By Proposition 1.3.3, $J(FG)$ is nilpotent, hence the image of $J(FG)$ under $\varepsilon_H : FG \to F\bar{G}$ is a nilpotent ideal of $F\bar{G}$. But the Jacobson radical of a ring contains every nil ideal. Thus, $(\bar{g}^k - 1)^2 \in \varepsilon_H(J(FG)) \subseteq J(F\bar{G})$. Of course, g^{k^l} is a p-element for each positive integer l. As $\bar{g}$ is a p-element, we can choose l in such a way that $\bar{g}^{k^l} = \bar{g}$. Thus, replacing k with k^l, we get $(\bar{g} - 1)^2 \in J(F\bar{G})$.

Let G be a group of smallest order satisfying the hypothesis of the lemma but not its conclusion. It must have two p-elements whose product is not a p-element and, by minimality, G must be generated by those two elements. Furthermore, we claim that G has an irreducible representation ρ of degree greater than 1. If not, then $FG/J(FG)$ is a direct sum of copies of F, and therefore commutative. It follows that for every $g \in G'$, $g - 1 \in J(FG)$. Now $J(FG)$ is nilpotent, hence g is a p-element. That is, G is commutative modulo a p-group and therefore, the p-elements of G form a subgroup.

Next, we claim that ρ is faithful. Suppose $1 \ne K = \ker(\rho)$. Surely $K \ne G$, so by minimality, the p-elements of K form a subgroup N, necessarily normal in G. If $N \ne 1$, then again by minimality of $|G|$, the p-elements of G/N form a subgroup as well. Thus, the p-elements of G form a subgroup, contradicting the choice of G. Therefore, K is a p'-group. We know that the p-elements of G/K form a normal subgroup L/K. But ρ induces a faithful irreducible representation on G/K. Since $\Delta(G/K, L/K)$ is nilpotent, by Lemma 1.1.1, it follows that

$\Delta(G/K, L/K) \subseteq J(F(G/K))$. That is, every element of L/K is in the kernel of every irreducible representation of G/K. Since this representation is faithful on G/K, $L = K$ and G has no p-elements, a contradiction. Thus, ρ is faithful on G.

Finally, if g is any p-element in G, then $(g-1)^2 \in J(FG)$, hence the minimal polynomial of $\rho(g)$ is either linear or quadratic. If it is linear, then $\rho(g)$ is a scalar multiple of the identity matrix. But g is a p-element, so this cannot happen in a field of characteristic p unless $g \in \ker(\rho)$, which is not allowed. Thus, the conditions of Proposition 2.3.3 are met, and G contains a subgroup isomorphic to $SL_2(\mathbb{Z}_p)$.

We know, therefore, that $SL_2(\mathbb{Z}_p)$ satisfies the hypothesis of the lemma, but by inspection, its p-elements do not form a subgroup. Thus, by minimality, $G = SL_2(\mathbb{Z}_p)$. Factoring out its centre, we obtain $PSL_2(\mathbb{Z}_p)$ and again, its p-elements do not form a subgroup. But this contradicts the minimality of $|G|$, and we are done. $\qquad \square$

Thus, if G is finite and $\mathcal{U}^+(FG)$ satisfies a group identity, then the p-elements of G form a normal subgroup P. In fact, we can simply consider G/P due to the following two results (which do not depend upon F being infinite).

Lemma 2.3.5. *Suppose that R is an F-algebra with involution, where char $F = p \neq 2$. Let I be a nil ideal invariant under $*$. If $\mathcal{U}^+(R)$ satisfies the group identity $w(x_1, \ldots, x_n) = 1$, then so does $\mathcal{U}^+(R/I)$. Conversely, if $p > 0$, I is nil of bounded exponent and $\mathcal{U}^+(R/I)$ satisfies a group identity, then $\mathcal{U}^+(R)$ satisfies a group identity.*

Proof. Suppose that $\mathcal{U}^+(R)$ satisfies a group identity $w(x_1, \ldots, x_n) = 1$. Choose symmetric units $\bar{u}_1, \ldots, \bar{u}_n \in \bar{R} = R/I$. By Lemma 2.3.1, we may assume that each $u_i \in \mathcal{U}^+(R)$. Thus, since $w(u_1, \ldots, u_n) = 1$, we must have $w(\bar{u}_1, \ldots, \bar{u}_n) = 1$.

Conversely, suppose that $p > 0$, I is nil of bounded exponent at most p^l and $\mathcal{U}^+(R/I)$ satisfies $w(x,y) = 1$. If $u, v \in \mathcal{U}^+(R)$, then $\bar{u}, \bar{v} \in \mathcal{U}^+(\bar{R})$, hence $w(\bar{u}, \bar{v}) = 1$. That is, $w(u,v) - 1 \in I$, hence $0 = (w(u,v) - 1)^{p^l} = w(u,v)^{p^l} - 1$. Thus, $\mathcal{U}^+(R/I)$ satisfies $(w(x,y))^{p^l} = 1$. $\qquad \square$

We now have the following lemma, similar to Lemma 1.2.18.

Lemma 2.3.6. *Let F be a field of characteristic $p > 2$ and G a group such that $\mathcal{U}^+(FG)$ satisfies a group identity $w(x_1, \ldots, x_n) = 1$. If N is a normal p-subgroup of G, and either N is finite or G is locally finite, then $\mathcal{U}^+(F(G/N))$ satisfies $w(x_1, \ldots, x_n) = 1$.*

Proof. If N is finite, then by Lemma 1.1.1, $\Delta(G,N)$ is nilpotent. It is surely $*$-invariant as well. By the preceding lemma, $\mathcal{U}^+(F(G/N))$ satisfies $w(x_1, \ldots, x_n) = 1$.

Now suppose that G is locally finite. Let $\bar{G} = G/N$. Take $\bar{\alpha}_1, \ldots, \bar{\alpha}_n \in \mathcal{U}^+(F\bar{G})$. We may lift the $\bar{\alpha}_i$ up to elements $\alpha_i \in (FG)^+$, and similarly for their inverses, since $(F\bar{G})^+$ is the image of $(FG)^+$ under $FG \to F\bar{G}$. Let H be the subgroup of G generated by the supports of all of these elements. Since H is finitely generated, it is finite. Therefore, since $\mathcal{U}^+(FH)$ satisfies $w(x_1, \ldots, x_n) = 1$, the finite case tells us that $\mathcal{U}^+(F(H/(H \cap N)))$ also satisfies $w(x_1, \ldots, x_n) = 1$. Replacing G with H and N with $H \cap N$, we obtain our result. $\qquad \square$

Thus, for finite groups G, $\mathscr{U}^+(FG)$ satisfies a group identity if and only if $\mathscr{U}^+(F(G/P))$ satisfies a group identity. Hence, we may assume that G has no elements with order divisible by char F. According to Lemma 2.2.5, every subgroup of G is normal, hence G is abelian or Hamiltonian. There is nothing to be said if G is abelian. In the Hamiltonian case, $G = Q_8 \times E \times O$, where E is an elementary abelian 2-group and O is abelian with every element having odd order. We would like to eliminate O. Then the finite case will be done, by Lemma 2.1.1. To this end, we have the following lemma.

Lemma 2.3.7. *Let F be an infinite field of characteristic $p > 2$. Let $G = Q_8 \times \langle c \rangle$. If $\mathscr{U}^+(FG)$ satisfies $w(x,y) = 1$, then there exists a positive integer m, depending only upon w, such that the order of c divides $2p^m$.*

Proof. Notice that the sets $\{\lambda^2 : \lambda \in \mathbb{Z}_p\}$ and $\{-1 - \mu^2 : \mu \in \mathbb{Z}_p\}$ each contain $\frac{p+1}{2}$ elements. Thus, they overlap, and we may choose $\lambda, \mu \in F$ such that $\lambda^2 + \mu^2 = -1$. Write $Q_8 = \langle g, h : g^2 = h^2, h^4 = 1, gh = h^{-1}g \rangle$. Then it is easily verified that $\theta : F(Q_8 \times \langle c \rangle) \to M_2(F\langle c \rangle)$ given by

$$\theta(g) = \begin{pmatrix} \lambda & \mu \\ \mu & -\lambda \end{pmatrix}, \quad \theta(h) = \begin{pmatrix} 0 & 1 \\ -1 & 0 \end{pmatrix}, \quad \text{and} \quad \theta(c) = \begin{pmatrix} c & 0 \\ 0 & c \end{pmatrix}$$

is a homomorphism.

It is also easy to check that $\alpha = \frac{1}{4}(c + c^{-1}g^2)(\mu g - h + \lambda gh)(1 - g^2)$ and $\beta = \frac{1}{4}(c + c^{-1}g^2)(\mu g + h + \lambda gh)(1 - g^2)$ are symmetric and square-zero. Thus, by Lemma 2.2.1, there exists an n, depending only on w, such that $(\alpha\beta)^n = 0$ and so $(\theta(\alpha)\theta(\beta))^n = 0$. But

$$\theta(\alpha) = \begin{pmatrix} 0 & c^{-1} - c \\ 0 & 0 \end{pmatrix} \quad \text{and} \quad \theta(\beta) = \begin{pmatrix} 0 & 0 \\ c^{-1} - c & 0 \end{pmatrix}.$$

Thus,

$$\begin{pmatrix} (c^{-1} - c)^{2n} & 0 \\ 0 & 0 \end{pmatrix} = \begin{pmatrix} 0 & 0 \\ 0 & 0 \end{pmatrix}.$$

That is, $(c^{-1} - c)^{2n} = 0$. Choosing m so that $p^m \geq 2n$, we get $c^{-p^m} = c^{p^m}$, hence $c^{2p^m} = 1$, as required. $\qquad\square$

Thus, if we have $G/P = Q_8 \times E \times O$, then as O cannot contain p-elements, we must have $O = 1$. As we mentioned above, this proves

Proposition 2.3.8. *Let F be an infinite field of characteristic $p > 2$ and G a finite group. Then $\mathscr{U}^+(FG)$ satisfies a group identity if and only if the p-elements of G form a (normal) subgroup P and G/P is abelian or a Hamiltonian 2-group.*

If char $F = 0$, we cannot use the technique of Lemma 2.3.7, since F may not contain λ and μ satisfying $\lambda^2 + \mu^2 = -1$. But we are still able to reduce to the Hamiltonian 2-groups. We will make use of the following famous result due to Tits (a corollary of Tits' alternative).

Proposition 2.3.9. *Suppose that $A, B \in GL_2(\mathbb{C})$, and that A has two eigenvalues with distinct magnitudes, as does B. Further suppose that the eigenspaces of A are distinct from those of B. Then there exists a positive integer n such that A^n and B^n generate a free group.*

Proof. See [95, Lemma 5.3]. □

We recall that for any field F of characteristic different from 2, the ring of (Hamiltonian) quaternions over F, $\mathbb{H}(F)$, is the F-vector space with basis $\{1, i, j, k\}$ made into an F-algebra through the rules $i^2 = j^2 = k^2 = -1$, $ij = -ji = k$, $jk = -kj = i$ and $ki = -ik = j$.

Lemma 2.3.10. *Let $G = Q_8 \times \langle c \rangle$, where $c \neq 1$ has odd order. Then $\mathscr{U}^+(\mathbb{Q}G)$ does not satisfy a group identity.*

Proof. We may assume that $o(c) = q$, an odd prime. It is easy to see that if ξ is a primitive qth root of unity, then the map $\mathbb{Q}\langle c \rangle \to \mathbb{Q}(\xi)$ given by $c \mapsto \xi$ is an epimorphism. As ξ has minimal polynomial $1 + x + \cdots + x^{q-1}$ over $\mathbb{Q}$, the kernel of this homomorphism is clearly $\mathbb{Q}\langle c \rangle (\frac{1}{q}\hat{c})$. Letting $e_1 = 1 - \frac{1}{q}\hat{c}$, we see that e_1 is a symmetric central idempotent, hence $\mathbb{Q}\langle c \rangle = \mathbb{Q}\langle c \rangle e_1 \oplus \mathbb{Q}\langle c \rangle (1 - e_1)$, and our epimorphism restricts to an isomorphism on $\mathbb{Q}\langle c \rangle e_1$.

Next, notice that the map $\mathbb{Q}Q_8 \to \mathbb{H}(\mathbb{Q})$ given by $g \mapsto i$, $h \mapsto j$ is also an epimorphism. Its kernel is easily seen to be $\mathbb{Q}Q_8(\frac{1+g^2}{2})$. Letting $e_2 = 1 - \frac{1+g^2}{2}$ we see that e_2 is a symmetric central idempotent and our epimorphism restricts to an isomorphism on $\mathbb{Q}Q_8 e_2$. Working in $\mathbb{Q}G \cong \mathbb{Q}Q_8 \otimes_{\mathbb{Q}} \mathbb{Q}\langle c \rangle$, we see that $e = e_1 e_2$ is a central symmetric idempotent and we have an epimorphism $\theta : \mathbb{Q}G \to \mathbb{H}(\mathbb{Q}(\xi))$ given by $g \mapsto i$, $h \mapsto j$, $c \mapsto \xi$, restricting to an isomorphism on $\mathbb{Q}Ge$.

Note that in $\mathbb{C}$, elements of the form $\lambda + \mu i$, with $\lambda, \mu \in \mathbb{Q}(\xi)$, are uniquely expressed in this form. Otherwise, we would get $i \in \mathbb{Q}(\xi)$. If this happened, then $\mathbb{Q}(\xi)$ would contain a primitive $4q$th root of unity. But the minimal polynomial of such a root of unity over $\mathbb{Q}$ has degree $\varphi(4q) = 2(q-1)$, where φ is the Euler function. However, $\varphi(q) = q - 1$, so this is impossible. Thus, we identify the elements of the form $\lambda + \mu i$ of $\mathbb{H}(\mathbb{Q}(\xi))$ with the subfield $\mathbb{Q}(\xi, i)$ of $\mathbb{C}$. (Equivalently, we could identify the elements $\lambda + \mu j$ with this subfield.)

Evidently, then, $(1 + \xi i)(1 - \xi^{-1}i)$ and $(1 + \xi j)(1 - \xi^{-1}j)$ are units in $\mathbb{H}(\mathbb{Q}(\xi))$. Furthermore,

$$(1 + \xi i)(1 - \xi^{-1}i) = \theta((1 + cg)(1 + (cg)^*)e)$$

and

$$(1 + \xi j)(1 - \xi^{-1}j) = \theta((1 + ch)(1 + (ch)^*)e).$$

As θ is an isomorphism on $\mathbb{Q}Ge$, $(1 + cg)(1 + (cg)^*)e$ and $(1 + ch)(1 + (ch)^*)e$ are units of $\mathbb{Q}Ge$; hence

$$(1 + cg)(1 + (cg)^*)e + (1 - e), (1 + ch)(1 + (ch)^*)e + (1 - e) \in \mathscr{U}^+(\mathbb{Q}G),$$

so $u = (1 + \xi i)(1 - \xi^{-1}i)$ and $v = (1 + \xi j)(1 - \xi^{-1}j)$ are images under θ of symmetric units in $\mathbb{Q}G$. It follows that $w(u, v) = 1$ if $w(x, y) = 1$ is a group identity for $\mathscr{U}^+(\mathbb{Q}G)$.

Now, $\mathbb{H}(\mathbb{Q}(\xi))$ may be regarded as a right $\mathbb{Q}(\xi, i)$-space with basis $\{1, j\}$. Consider the left regular representation on $\mathbb{H}(\mathbb{Q}(\xi))$; that is, the map sending r to ρ_r, where $\rho_r(s) = rs$ for all s. Surely $w(\rho_u, \rho_v) = 1$. But with respect to the basis $\{1, j\}$ we calculate the matrix of ρ_u to be

$$A = \begin{pmatrix} 2 + (\xi - \xi^{-1})i & 0 \\ 0 & 2 + (\xi^{-1} - \xi)i \end{pmatrix}$$

and the matrix of ρ_v to be

$$B = \begin{pmatrix} 2 & \xi^{-1} - \xi \\ \xi - \xi^{-1} & 2 \end{pmatrix}.$$

The eigenvalues of A are clearly distinct positive real numbers. A calculation reveals that the eigenvalues of B are the same as those of A. Furthermore, the eigenvectors of A are obvious and are not eigenvectors for B. Thus, the preceding proposition applies, and we get that powers of A and B generate a free group. This contradicts the hypothesis that they satisfy a group identity, and we are done. $\qquad\square$

We can now conclude the torsion case for fields of characteristic zero.

Proposition 2.3.11. *Let F be a field of characteristic zero and G a torsion group. Then $\mathscr{U}^+(FG)$ satisfies a group identity if and only if G is abelian or a Hamiltonian 2-group.*

Proof. By Lemma 2.2.5, if $\mathscr{U}^+(FG)$ satisfies a group identity, then G is abelian or Hamiltonian. In the latter case, $G \simeq Q_8 \times E \times O$, where $E^2 = 1$ and every element of O has odd order. By the last lemma, $O = 1$. The converse follows immediately from Lemma 2.1.1. $\qquad\square$

2.4 Group Rings of Torsion Groups

Having dealt with the characteristic zero case, we can now assume that F is infinite and char $F = p > 2$. The next step will be to show that if $\mathscr{U}^+(FG)$ satisfies a group identity, then FG satisfies a polynomial identity. This breaks down into three cases: $N(FG) = 0$, $N(FG)$ nonzero nilpotent, and $N(FG)$ not nilpotent. In the first of these cases, where FG is semiprime, we can do even better. We start with

Lemma 2.4.1. *Suppose F is an infinite field, char $F = p > 2$ and FG is semiprime. Further suppose that $\mathscr{U}^+(FG)$ satisfies a group identity. If $g \in G$ has order p, then $\langle g \rangle$ is normalized by every torsion element of G.*

Proof. Take $h \in G$ with $o(h) = p^m$, $m \geq 0$. Observe that $\hat{g}$ is symmetric and square-zero and $(h - 1)(h^{-1} - 1)$ is symmetric with $((h - 1)(h^{-1} - 1))^{p^m} = 0$. Thus, by Lemma 2.2.7, $\hat{g}(h - 1)(h^{-1} - 1)\hat{g} = 0$. Expanding this and discarding the $(\hat{g})^2$ terms, we get $\hat{g}h\hat{g} + \hat{g}h^{-1}\hat{g} = 0$. Writing this as a sum of group elements, we note that each element must appear at least three times in order to get a zero sum. Thus, a group element appears at least twice in either $\hat{g}h\hat{g}$ or $\hat{g}h^{-1}\hat{g}$. As the proofs for the two cases are essentially the same, assume that $g^{i_1}hg^{j_1} = g^{i_2}hg^{j_2}$. If $i_1 = i_2$, then $j_1 = j_2$, so we may assume that $i_1 \neq i_2$ and $j_1 \neq j_2$, with all exponents in the range of 0 to $p-1$. Then $h^{-1}g^{i_1-i_2}h = g^{j_2-j_1}$. As $i_1 - i_2$ is not divisible by p, by taking a suitable power we get that $h^{-1}gh \in \langle g \rangle$. Thus, $\langle g \rangle$ is normalized by the p-elements of G.

Suppose that $k \in G$ is a p'-element. By Lemma 2.2.5, $\langle k \rangle$ is normal in G. Thus, $\langle k, g \rangle$ is finite. We can now apply Proposition 2.3.8 to $\langle k, g \rangle$, and we get that its p-elements form a normal subgroup. Therefore, (g,k) is both a p-element and a p'-element, hence g and k commute. Thus, $\langle g \rangle$ is normalized by both the p-elements and the p'-elements and therefore by all elements of finite order. $\square$

Proposition 2.4.2. *Suppose F is an infinite field, char $F \neq 2$, G is torsion and FG is semiprime. Then $\mathscr{U}^+(FG)$ satisfies a group identity if and only if G is abelian or a Hamiltonian 2-group.*

Proof. The characteristic zero case is handled in Proposition 2.3.11, so let char $F = p > 2$ and suppose that $\mathscr{U}^+(FG)$ satisfies a group identity. By the preceding lemma, if $g \in G$ has order p, then $\langle g \rangle$ is a normal subgroup. But by Proposition 1.2.9, this contradicts the semiprimeness of FG. Thus, G has no p-elements. By Lemma 2.2.5, G is abelian or Hamiltonian. In the latter case, $G = Q_8 \times E \times O$, where $E^2 = 1$ and O is an abelian group in which every element has odd, p'-order. By Lemma 2.3.7, $O = 1$, and the necessity is proved.

Lemma 2.1.1 proves the sufficiency. $\square$

Let us now extend Hartley's conjecture.

Proposition 2.4.3. *Let F be an infinite field of characteristic $p \neq 2$ and G a torsion group. If $\mathscr{U}^+(FG)$ satisfies a group identity, then FG satisfies a polynomial identity. In particular, G is locally finite. Also, if $p > 2$, then the p-elements of G form a (normal) subgroup.*

Proof. Once we show that FG satisfies a polynomial identity, the fact that G is locally finite will follow from Proposition 1.1.4, and then the fact that the p-elements form a subgroup will come from Proposition 2.3.8. If $p = 0$, then Propositions 2.3.11 and 1.1.4 do the job. Thus, we assume that $p > 2$.

Suppose, first of all, that $N(FG)$ is a nilpotent ideal. By Proposition 1.2.22, $\phi_p(G)$ is finite, and by Proposition 2.3.8, its p-elements form a group. Thus, $\phi_p(G)$ is a finite p-group. By Lemma 2.3.6, $\mathscr{U}^+(F(G/\phi_p(G)))$ satisfies a group identity. But by Proposition 1.2.9, $F(G/\phi_p(G))$ is semiprime. Thus, by Proposition 2.4.2, $G/\phi_p(G)$ is abelian or a Hamiltonian 2-group. If it is abelian, then $[FG, FG] \subseteq \Delta(G, \phi_p(G))$. But $\Delta(G, \phi_p(G))$ is a nilpotent ideal by Lemma 1.1.1, hence FG satisfies $[x,y]^{p^l} = 0$ for some l. If, on the other hand, $G/\phi_p(G)$ is a Hamiltonian 2-group, then by Lemma 2.1.1, $(F(G/\phi_p(G)))^+$ is commutative. That is, $[(FG)^+, (FG)^+] \subseteq \Delta(G, \phi_p(G))$, hence FG satisfies the $*$-polynomial identity $[x+x^*, y+y^*]^{p^l} = 0$. By Proposition 2.1.2, FG satisfies a polynomial identity.

Now, suppose that $N(FG)$ is not nilpotent. It is surely nil, so by Lemma 2.2.6, $N(FG)$ satisfies a polynomial identity. By Lemma 1.2.16, FG satisfies a nondegenerate multilinear GPI. Thus, by Proposition 1.2.15, $(G : \phi(G)) < \infty$ and $|(\phi(G))'| < \infty$. In particular, G is locally finite. Considering finite subgroups of G, we see from Proposition 2.3.8 that the p-elements of G form a normal subgroup P.

By Lemma 2.3.6, $\mathcal{U}^+(F(G/P))$ satisfies a group identity. But $F(G/P)$ is semiprime, hence by Proposition 2.4.2, G/P is abelian or a Hamiltonian 2-group. In the former case, G' is a p-group, hence $(\phi(G))'$ is a finite p-group. That is, G has a p-abelian subgroup of finite index, and by Proposition 1.1.4, FG satisfies a polynomial identity. If $G/P \simeq Q_8 \times E$, where E is an elementary abelian 2-group, then there exists a normal subgroup H of G, containing P, such that $G/H \simeq Q_8$ and $H/P \simeq E$. Since H/P is abelian, as we have just seen, FH satisfies a polynomial identity. That is, H has a p-abelian subgroup of finite index and, therefore, so does G.   □

As usual, we write P for the subgroup of G consisting of p-elements. Let us first consider the case where G does not contain the quaternions. Let H be any finitely generated (hence finite) subgroup of G. Let N be the group of p-elements of H. By Proposition 2.3.8, H/N is abelian or a Hamiltonian 2-group. Now, $(|N|, |H/N|) = 1$. Thus, by the Schur–Zassenhaus theorem, H is the semidirect product $N \rtimes K$, for some subgroup K. If H/N is a Hamiltonian 2-group, then K contains the quaternions, which is impossible. So H/N is abelian, and H', and hence G', is a p-group. We already know that FG satisfies a polynomial identity. In view of Theorem 1.3.1, if we can show that G' has bounded exponent, then we will know that $\mathcal{U}(FG)$ satisfies a group identity and we will be done. In order to do this, we need versions of Lemmas 1.3.5 and 1.3.6, assuming only that $\mathcal{U}^+(FG)$ satisfies a group identity (and that F is infinite).

Lemma 2.4.4. *Let F be an infinite field of characteristic $p > 2$ and let $G = A \rtimes \langle g \rangle$, where A is an abelian p-subgroup and g has prime order $q \neq p$. If $\mathcal{U}^+(FG)$ satisfies a group identity, then G' has bounded exponent.*

Proof. Define $\theta : FA \to FA$ as in the proof of Lemma 1.3.5, and recall that for all $\beta \in FA$, $\theta(\beta)$ is central and $\hat{g}\beta\hat{g} = \theta(\beta)\hat{g}$. Take any $a \in A$. Let $\alpha = \hat{g}a^{-1}(1 - g^{-1})$. Clearly $\alpha^2 = 0$; hence, by Lemma 2.2.1, there exists an n (independent of the choice of a) such that $(\alpha\alpha^*)^{p^n} = 0$. But $\alpha^* = (1 - g)a\hat{g}$. Thus,

$$
\begin{aligned}
\alpha\alpha^* &= \hat{g}a^{-1}(2 - g - g^{-1})a\hat{g} \\
&= 2(\hat{g})^2 - \hat{g}a^{-1}ga\hat{g} - \hat{g}a^{-1}g^{-1}a\hat{g} \\
&= 2q\hat{g} - \hat{g}(g,a)\hat{g} - \hat{g}(g^{-1},a)\hat{g} \\
&= (2q - \theta((g,a)) - \theta((g^{-1},a)))\hat{g}.
\end{aligned}
$$

Now, $2q - \theta((g,a)) - \theta((g^{-1},a))$ is central, so it follows that

$$
(2q - \theta((g,a)) - \theta((g^{-1},a)))^{p^n} q^{p^n-1}\hat{g} = 0.
$$

As $2q - \theta((g,a)) - \theta((g^{-1},a)) \in FA$, it follows that

$$(2q - \theta((g,a)) - \theta((g^{-1},a)))^{p^n} = 0.$$

Everything in FA commutes, hence

$$(2q)^{p^n} = \theta((g,a))^{p^n} + \theta((g^{-1},a))^{p^n}.$$

As $(2q)^{p^n}$ is a nonzero multiple of the identity element, one of the conjugates of (g,a) or (g^{-1},a) appearing in this last equation must have order dividing p^n. Thus, (g,a) or (g^{-1},a) has order dividing p^n, and by Lemma 1.3.4, G' has bounded exponent.   □

Lemma 2.4.5. *Let F be an infinite field of characteristic $p > 2$ and let $G = A \rtimes \langle g \rangle$, where A is an abelian p-subgroup and g has order p. If $\mathscr{U}^+(FG)$ satisfies a group identity, then G' has bounded exponent.*

Proof. Notice that $\hat{g}$ is square-zero and symmetric and, for any $a \in A$, so is $a^{-1}\hat{g}a$. Hence, by Lemma 2.2.1, there exists r, independent of the choice of a, such that $(a^{-1}\hat{g}a\hat{g})^{p^r} = 0$. Now follow the proof of Lemma 1.3.6.   □

We also need the following lemma.

Lemma 2.4.6. *Let F be any field and G a group. Let N be a torsion normal subgroup of G having no elements of order divisible by char F. Suppose either that N is finite or G is locally finite. If $\mathscr{U}^+(FG)$ satisfies the group identity $w(x_1,\ldots,x_n) = 1$, then so does $\mathscr{U}^+(F(G/N))$.*

Proof. In the proof of Lemma 1.3.9, we simply note that e is symmetric, and now apply the same proof to the symmetric units.   □

Remark 2.4.7. Notice that if F is infinite and char $F \neq 2$, then by Proposition 2.4.3, if G is torsion and $\mathscr{U}^+(FG)$ satisfies a group identity, then we get for free that G is locally finite. If char $F = 0$, this implies that we can choose any normal subgroup N in the above lemma. If char $F = p > 2$, then we know that the p-elements of N form a normal subgroup, so we can factor out the p-elements and then the p'-elements. Thus, we can factor out any normal subgroup N here as well. This is obviously the case for group identities on $\mathscr{U}(FG)$ too (even if char $F = 2$).

The first part of the main result of Giambruno et al. [39] is

Theorem 2.4.8. *Let F be an infinite field of characteristic $p \neq 2$ and G a torsion group not containing Q_8. Then the following are equivalent:*

(i) $\mathscr{U}^+(FG)$ satisfies a group identity;
(ii) $\mathscr{U}(FG)$ satisfies a group identity;
(iii)   a. $p = 0$ and G is abelian, or
*   b. $p > 2$, G has a p-abelian subgroup of finite index and G' is a p-group of bounded exponent.*

Proof. It is clear that (ii) implies (i), and (iii) implies (ii) comes from Theorem 1.3.1 (or is trivial if $p = 0$). Thus, we must show that (i) implies (iii). The characteristic zero case was dealt with in Proposition 2.3.11. Thus, let p be an odd prime, and assume that $\mathscr{U}^+(FG)$ satisfies a group identity. We have already seen that FG satisfies a polynomial identity (hence G has a p-abelian normal subgroup A of finite index, by Proposition 1.1.4) and G' is a p-group. Thus, we need only show that G' has bounded exponent.

By Lemma 2.3.6, $\mathscr{U}^+(F(G/A'))$ satisfies a group identity. It suffices to show that $(G/A')'$ has bounded exponent. Thus, we factor out A' and assume that A is abelian. Write $A = H \times K$, where H is a p-group and K is a p'-group. Then K is normal and by Lemma 2.4.6, $\mathscr{U}^+(F(G/K))$ satisfies a group identity. Also, $(G/K)' = G'K/K \simeq G'/(G' \cap K) = G'$, since G' is a p-group. Thus, we factor out K and assume that A is a p-group. In view of Lemmas 2.4.4 and 2.4.5, the proof of Lemma 1.3.10 shows us that G' has bounded exponent. We are done. □

If G contains the quaternions, then $\mathscr{U}(FG)$ cannot satisfy a group identity for any infinite field of characteristic different from 2 (see Theorem 1.3.2 and Corollary 1.2.21). However, the second main theorem from [39] is

Theorem 2.4.9. *Let F be an infinite field of characteristic $p \neq 2$ and G a torsion group containing Q_8. If $p = 0$, then $\mathscr{U}^+(FG)$ satisfies a group identity if and only if G is a Hamiltonian 2-group. If $p > 2$, then $\mathscr{U}^+(FG)$ satisfies a group identity if and only if G has a p-abelian normal subgroup A of finite index, the p-elements of G form a (normal) subgroup P of bounded exponent, and G/P is a Hamiltonian 2-group.*

Proof. The characteristic zero case was resolved in Proposition 2.3.11, so suppose $p > 2$. Assume that $\mathscr{U}^+(FG)$ satisfies a group identity. By Proposition 2.4.3, FG satisfies a polynomial identity, and Proposition 1.1.4 shows the existence of A. Proposition 2.4.3 also says that P is a subgroup. As G is locally finite, Lemma 2.3.6 tells us that $\mathscr{U}^+(F(G/P))$ satisfies a group identity. But $F(G/P)$ is semiprime. Thus, by Proposition 2.4.2, G/P is abelian or a Hamiltonian 2-group. It surely cannot be abelian, since G contains Q_8.

Thus, it remains only to show that P has bounded exponent. By Theorem 2.4.8, P' has bounded exponent. Therefore, it suffices to show that P/P' has bounded exponent. We know that we can factor out P'. Thus, we assume that P is abelian. Suppose that $G/P = (HL)/P$ where $H/P \simeq Q_8$ and L/P is an elementary abelian 2-group. Take any $g \in H$. Then $\langle P, g \rangle$ does not contain Q_8. Thus, by Theorem 2.4.8, $\langle P, g \rangle'$ has bounded exponent. Similarly, $\langle P, L \rangle'$ has bounded exponent. Noting that in any group we have $(gh, k) = (g, k)^h (h, k)$, we see that since P is abelian, it follows that (P, G) has bounded exponent. Thus, we can safely quotient out (P, G) and assume that P is central. But then G has a subgroup isomorphic to $Q_8 \times P$. It follows from Lemma 2.3.7 that P has bounded exponent, and the necessity is complete.

Let us prove the sufficiency. But FG satisfies a polynomial identity and P is a normal p-subgroup of bounded exponent. Thus, by Lemma 1.3.14, $\Delta(G, P)$ is nil of bounded exponent at most p^k, for some k. Now, if $\alpha, \beta \in \mathscr{U}^+(FG)$, then by

Lemma 2.1.1, $(\alpha,\beta) - 1 \in \Delta(G,P)$ and therefore, $(\alpha,\beta)^{p^k} = 1$. Thus, $\mathscr{U}^+(FG)$ satisfies a group identity. $\square$

Combining Theorems 2.4.8 and 1.3.1 with the proof of Theorem 2.4.9, we obtain the following fact. If F is an infinite field of characteristic $p \neq 2$ and G is a torsion group, and if $\mathscr{U}^+(FG)$ satisfies a group identity, then it satisfies $(x,y)^{p^k} = 1$ for some $k \geq 0$ (or $(x,y) = 1$ if $p = 0$).

2.5 Semiprime Group Rings

Let us move on to the results of Sehgal and Valenti [96] determining when $\mathscr{U}^+(FG)$ satisfies a group identity if G has elements of infinite order. Once again, we let F be an infinite field of characteristic $p \neq 2$. This time, G will be an arbitrary group. We let T denote the set of torsion elements, P the set of p-elements, and Q the set of p'-elements in T. (For convenience, when $p = 0$, we let $P = 1$ and $Q = T$.)

We begin with the semiprime case. By Lemma 2.2.5, if $g \in Q$, then $\langle g \rangle$ is a normal subgroup. It follows that Q is a normal subgroup (and, indeed, every subgroup of Q is normal in G). Thus, by Proposition 2.4.2, Q is abelian or a Hamiltonian 2-group. Curiously, this will be the only time we have to worry about the quaternions; indeed, they will only appear when the characteristic is zero. Let us begin with

Lemma 2.5.1. *Let F be an infinite field of characteristic different from 2. Suppose that G has a finite subgroup H and an element g of infinite order such that $G = \langle H,g \rangle$. Further suppose that FG is semiprime. If $\mathscr{U}^+(FG)$ satisfies a group identity, and FH has no nilpotent elements, then every idempotent of FH is central in FG.*

Proof. By Proposition 1.3.3, $J(FH)$ is nilpotent. Thus, by our assumption on FH, $J(FH) = 0$. Therefore, FH is a direct sum of matrix rings over division rings. But any 2×2 or larger matrix ring surely has nilpotent elements. Therefore, FH is a direct sum of division rings. As division rings only have 0 and 1 as idempotents, it follows that the idempotents of FH are the sums of the primitive central idempotents. Thus, it suffices to show that the primitive central idempotents of FH are central in FG.

Let e be a primitive central idempotent of FH. By Lemma 2.2.4, any symmetric idempotent is central in FG. Thus, assume $e^* \neq e$. Now, e^* is also a primitive central idempotent of FH. Since $e + e^*$ is a symmetric idempotent, it is central in FG. That is, $e^g + (e^*)^g = e + e^*$. As we observed above, H is normal in G. Thus, e^g and $(e^*)^g$ are also primitive central idempotents of FH, and we must have $e^g = e$ or e^*. In the former case, e is centralized by H and g and is therefore central in FG. That is, we may assume that $e^g = e^*$. Repeating the same procedure with g^{-1} in place of g, we get $e^{g^{-1}} = e^*$. Expanding these equations, we get $eg = ge^*$, $g^{-1}e = e^*g^{-1}$, $eg^{-1} = g^{-1}e^*$ and $ge = e^*g$. Let $\alpha = (g + g^{-1})e$ and $\beta = (g + g^{-1})e^*$. Now,

$$\alpha^* = e^*(g + g^{-1}) = (g + g^{-1})e = \alpha$$

and similarly, $\beta^* = \beta$. Also,

$$\alpha^2 = (g + g^{-1})e(g + g^{-1})e = (g + g^{-1})^2 e^* e = 0,$$

as e and e^* are distinct primitive central idempotents. Similarly, $\beta^2 = 0$. By Lemma 2.2.1, $\alpha\beta$ is nilpotent. But $\alpha\beta = (g + g^{-1})^2 (e^*)^2 = (g + g^{-1})^2 e^*$. It follows that for any n, $(\alpha\beta)^n = (g + g^{-1})^{2n} e^*$. If this is 0, then so is $g^{2n}(g + g^{-1})^{2n} e^*$. Expanding this last expression, we get e^* plus a linear combination of group elements outside of H. Thus, $(\alpha\beta)^n \neq 0$, and we have a contradiction. $\qquad\square$

We intend to apply this lemma when H is a Hamiltonian 2-group. Happily, the group rings of Q_8 have been studied thoroughly. The following result is classical.

Proposition 2.5.2. *Let F be any field. Then FQ_8 has a nonzero nilpotent element if and only if there exist $\lambda, \mu \in F$ such that $\lambda^2 + \mu^2 = -1$.*

Proof. See [94, Proposition VI.1.13]. $\qquad\square$

Lemma 2.5.3. *Let F be an infinite field of characteristic different from 2 and G a group generated by a finite Hamiltonian 2-group H and an element a of infinite order. If FG is semiprime and $\mathcal{U}^+(FG)$ satisfies a group identity, then char $F = 0$ and every idempotent of FH is central in FG.*

Proof. Suppose, first of all, that $\lambda^2 + \mu^2 = -1$ has no solution in F. By Proposition 2.5.2, FQ_8 has no nilpotent elements. Now, $H = Q_8 \times E$, where $E = C_2 \times \cdots \times C_2$. Thus, $FH = FQ_8 \otimes_F FC_2 \otimes_F FC_2 \otimes_F \cdots \otimes_F FC_2$. But if $C_2 = \langle c \rangle$, then

$$FC_2 = FC_2 \left(\frac{1+c}{2} \right) \oplus FC_2 \left(\frac{1-c}{2} \right) \cong F \oplus F.$$

Thus, FH is a direct sum of copies of FQ_8 and therefore has no nilpotent elements. It follows from Lemma 2.5.1 that every idempotent of FH is central in FG. Also, we know that $\lambda^2 + \mu^2 = -1$ has a solution in $\mathbb{Z}_p$ for every prime p. Thus, char $F = 0$.

It remains to dispense with the case where $\lambda^2 + \mu^2 = -1$ has a solution. We know that every subgroup of H is normal in G. Thus, conjugation by a induces an automorphism of Q_8 under which every subgroup is invariant. By inspection, we can see that any such automorphism must be conjugation by an element of Q_8, say k. But then ak^{-1} centralizes Q_8. Furthermore, $G = \langle H, ak^{-1} \rangle$ and since H is finite and normal, this implies that ak^{-1} has infinite order. Thus, we replace a with ak^{-1} and assume that $G = Q_8 \times \langle a \rangle$.

Writing $Q_8 = \langle g, h \rangle$, we know that there is an epimorphism $\theta : FQ_8 \to M_2(F)$ given by

$$\theta(g) = \begin{pmatrix} \lambda & \mu \\ \mu & -\lambda \end{pmatrix} \quad \text{and} \quad \theta(h) = \begin{pmatrix} 0 & 1 \\ -1 & 0 \end{pmatrix}.$$

Indeed, an easy calculation reveals that the matrices $\theta(g)$, $\theta(h)$ and $\theta(gh)$, together with the identity matrix, form a basis of $M_2(F)$. In fact, the kernel of θ is $FQ_8(\frac{1+g^2}{2})$, which is clearly an ideal invariant under $*$, as g^2 is both central and symmetric, and $FQ_8(\frac{1-g^2}{2}) \cong M_2(F)$. Thus, θ induces an involution on $M_2(F)$ and furthermore, if $\theta(\gamma)$ is a symmetric unit in $M_2(F)$, then $\gamma(\frac{1-g^2}{2}) + \frac{1+g^2}{2}$ is a symmetric unit in FQ_8. This means that the symmetric units in $M_2(F)$ are images of symmetric units in FQ_8.

Observe that the traces of $\theta(g)$, $\theta(h)$ and $\theta(gh)$ are all zero. Furthermore, $(\theta(g))^* = \theta(g^*) = \theta(g^{-1}) = -\theta(g)$, and similarly for h and gh. Thus, $\theta(g)$, $\theta(h)$ and $\theta(gh)$ span a three-dimensional subspace of the set of skew elements in $M_2(F)$. Surely the identity matrix is symmetric, so $(M_2(F))^- \neq M_2(F)$. Thus, these three matrices are a basis for $(M_2(F))^-$. Similarly, they are a basis for the set of matrices of trace zero, so the matrices of trace zero are precisely the skew matrices.

We can extend θ to $\theta : FG \to M_2(F\langle a \rangle)$ via

$$\theta(a) = \begin{pmatrix} a & 0 \\ 0 & a \end{pmatrix}.$$

Thus, the symmetric units in $M_2(F\langle a \rangle)$ are images of symmetric units in FG, and the symmetric units in $M_2(F\langle a \rangle)$ satisfy a group identity.

Now, E_{12} and E_{21} are both skew, having trace zero. Furthermore,

$$A = \begin{pmatrix} a - a^{-1} & 0 \\ 0 & a - a^{-1} \end{pmatrix}$$

is central and obviously skew, by definition of θ. Therefore,

$$AE_{12} = \begin{pmatrix} 0 & a - a^{-1} \\ 0 & 0 \end{pmatrix} \quad \text{and} \quad AE_{21} = \begin{pmatrix} 0 & 0 \\ a - a^{-1} & 0 \end{pmatrix}$$

are both symmetric and square-zero. But their product is

$$\begin{pmatrix} (a - a^{-1})^2 & 0 \\ 0 & 0 \end{pmatrix},$$

and this is not nilpotent. By Lemma 2.2.1, we have a contradiction.   □

Suppose that FG is semiprime and $\mathcal{U}^+(FG)$ satisfies a group identity. We know that Q is abelian or a Hamiltonian 2-group. Choose any idempotent $e \in FQ$. Let $a \in G$ have infinite order and take any $b \in Q$. Since Q is locally finite, we can find a finite subgroup H of Q containing b and the support of e. We know that H is normal in G. Thus, in $\langle H, a \rangle$, every element outside of H has infinite order, so $F\langle H, a \rangle$ is semiprime. Once again, H is abelian or a Hamiltonian 2-group. In the former case, FH is a direct sum of fields, and therefore Lemma 2.5.1 applies. In the latter case, Lemma 2.5.3 applies. Either way, e commutes with a and b. That is, e commutes with Q and with every element of infinite order. When char $F = 0$, this implies that e is central in FG. If char $F = p > 2$, then to make the same statement, we must prove that $Q = T$; that is, that G has no p-elements.

Lemma 2.5.4. *Let F be an infinite field of characteristic $p > 2$. Suppose that G has an element of infinite order and FG is semiprime. If $\mathcal{U}^+(FG)$ satisfies a group identity, then G has no p-elements.*

Proof. Let $g \in P$ have order p. By Lemma 2.4.1, $\langle g \rangle$ is normalized by T. Thus, the elements of order p in G generate an elementary abelian p-subgroup H, which is obviously normal in G. Let $a \in G$ have infinite order. We claim that the set $\{g^{a^i} : i \geq 0\}$ is finite.

Suppose this is not the case. If $g^a \in \langle g \rangle$, then $g^{a^i} \in \langle g \rangle$ for all $i \geq 0$, contradicting our hypothesis. So, we may assume that $g^a \notin \langle g \rangle$. As H is an elementary abelian p-group, we have a direct product $\langle g \rangle \times \langle g^a \rangle$. Similarly, we may assume that $g^{a^2} \notin \langle g, g^a \rangle$ and $g^{a^3} \notin \langle g, g^a, g^{a^2} \rangle$, and we get a direct product $\langle g \rangle \times \langle g^a \rangle \times \langle g^{a^2} \rangle \times \langle g^{a^3} \rangle$. Noting that $\hat{g}(a + a^{-1})\hat{g}$ is square-zero and symmetric and that $(\hat{g})^{a^2}(1 - g^{a^2}) = 0$, we see from Lemma 2.2.7 that

$$(\hat{g})^{a^2}\hat{g}(a + a^{-1})\hat{g}(1 - g^{a^2}) = 0.$$

Now, H is normal in G and $g \in H$, but $a^2 \notin H$, hence a and a^{-1} lie in different cosets modulo H. Thus, the linear combinations of elements in Ha must sum to zero, as must those in Ha^{-1}. That is,

$$(\hat{g})^{a^2}\hat{g}a^{-1}\hat{g}(1 - g^{a^2}) = 0.$$

Simplifying and multiplying on the right by a, we obtain

$$(\hat{g})^{a^2}\hat{g}(\hat{g})^a(1 - g^{a^3}) = 0.$$

But we have zero equal to a product of terms lying, respectively, in $F\langle g^{a^2} \rangle$, $F\langle g \rangle$, $F\langle g^a \rangle$ and $F\langle g^{a^3} \rangle$. As the product of the groups in question is direct, one of these terms is zero, which is not the case. The claim is proved.

Let $K = \langle g, a \rangle$, and let N be the subgroup of K generated by the finite set of conjugates of g discussed above. Then N is normal in K and since $N \leq H$, N is a finite elementary abelian p-group. Thus, by Lemma 1.1.1, $\Delta(K,N)$ is nilpotent. Hence, for any group element g of order p and any element a of infinite order, we get that $(1 - g)a^{-1} + a(1 - g^{-1})$ is nilpotent and symmetric. By Proposition 1.2.9, $\langle g \rangle$ is not a normal subgroup of G. Thus, there exists a conjugate g' of g with $g' \notin \langle g \rangle$. As g' has order p as well, we see that $\hat{g}'$ is symmetric and square-zero. Thus, by Lemma 2.2.7,

$$\hat{g}'((1 - g)a^{-1} + a(1 - g^{-1}))\hat{g}' = 0.$$

Once again, a and a^{-1} do not lie in the same coset of H, so separating the terms corresponding to each coset, we see that $\hat{g}'(1 - g)a^{-1}\hat{g}' = 0$. Multiplying by a, we obtain

$$\hat{g}'(1 - g)(\hat{g}')^a = 0.$$

If $(g')^a \notin \langle g \rangle \times \langle g' \rangle$, then we have a direct product $\langle g \rangle \times \langle g' \rangle \times \langle (g')^a \rangle$, in which case one of the terms in the above product is zero, which is not the case. Thus, $(g')^a \in \langle g, g' \rangle$. Similarly, $g^a \in \langle g, g' \rangle$. That is, $\langle g, g' \rangle$ is a subgroup of order p^2 normalized by every element of infinite order. We saw above that it is normalized by T as well; hence, the subgroup is normal. But this contradicts Proposition 1.2.9. Thus, g does not exist, and G has no p-elements. $\qquad\square$

In summary, we know that $T = Q$, and T is abelian or a Hamiltonian 2-group (with the latter case only occurring if the characteristic is zero). Also, every idempotent in FT is central in FG. One last condition is required.

Lemma 2.5.5. *Let F be an infinite field of characteristic different from 2 and G a group such that FG is semiprime. If FT has a nonsymmetric idempotent and $\mathcal{U}^+(FG)$ satisfies a group identity, then G/T satisfies a group identity.*

Proof. If G is torsion, there is nothing to say, so let G have an element of infinite order and suppose that e is a nonsymmetric idempotent of FT. If $\mathcal{U}^+(FG)$ satisfies $w(x, y) = 1$, then we claim that G/T satisfies this identity as well. Let H be the (finite normal) subgroup of T generated by the support of e. Since e is known to be central, it is a sum of primitive central idempotents of FH. Thus, since e is not symmetric, one of the primitive central idempotents is not symmetric. Say

$$FH = FHe_1 \oplus FHe_1^* \oplus FHe_3 \oplus \cdots \oplus FHe_n$$

is the Wedderburn decomposition of FH. Since each idempotent of FH is central, we have

$$FG = FGe_1 \oplus FGe_1^* \oplus FGe_3 \oplus \cdots \oplus FGe_n$$

as well. If $g \in G$, then $ge_1 + g^{-1}e_1^* + e_3 + \cdots + e_n$ is a symmetric unit. Thus, for any $g, h \in G$,

$$w(ge_1 + g^{-1}e_1^* + e_3 + \cdots + e_n, he_1 + h^{-1}e_1^* + e_3 + \cdots + e_n) = 1.$$

Looking only at the first component, we have $w(ge_1, he_1) = e_1$, hence $w(g, h)e_1 = e_1$. But $w(g, h)$ is a group element, and if $w(g, h)e_1 = e_1$, then $w(g, h)$ lies in H. That is, G/T satisfies $w(x, y) = 1$. $\qquad\square$

The first part of the main result of [96] is

Theorem 2.5.6. *Let F be an infinite field of characteristic $p \neq 2$, and let G be a group containing an element of infinite order. Suppose that FG is semiprime. If $\mathcal{U}^+(FG)$ satisfies a group identity, then*

1. *if $p = 0$, then the set of torsion elements, T, is a (normal) subgroup of G, and T is abelian or a Hamiltonian 2-group;*
2. *if $p > 2$, then T is an abelian (normal) p'-subgroup of G;*
3. *every idempotent in FT is central in FG; and*
4. *if FT contains a nonsymmetric idempotent, then G/T satisfies a group identity.*

Conversely, if G/T is a u.p. group and FG satisfies the above four conditions, then $\mathscr{U}^+(FG)$ satisfies a group identity.

Proof. We have only to verify the converse. Of course we are done if the conditions of Theorem 1.4.9 are satisfied, for in this case $\mathscr{U}(FG)$ satisfies a group identity. First of all, note that in the proof of Theorem 1.4.9 it is sufficient to assume for the third condition that every idempotent of FT is central in FG. For the remaining conditions, there are only two cases to consider, namely, if $p = 0$ and T is a Hamiltonian 2-group, or if every idempotent of FT is symmetric. In fact, the former case is contained in the latter. We know that every idempotent of FT is central and, by Lemma 2.1.1, every central element in FT, where T is a Hamiltonian 2-group, is symmetric. Thus, in any case, we can assume that every idempotent of FT is symmetric. We claim that $\mathscr{U}^+(FG)$ is abelian.

Take any $\alpha \in \mathscr{U}^+(FG)$. By Remark 1.4.10, there is a finite subgroup E of T such that for any primitive idempotent e of FE, we have $\alpha e = \lambda g$ for some $\lambda \in \mathscr{U}(FEe)$ and some $g \in G$. Since e is symmetric and central, $\lambda g = (\lambda g)^* = g^{-1}\lambda^*$. Thus, $g^2 e = \lambda^{-1}(g^{-1}\lambda^* g) \in FEe$. (Remember that E is a normal subgroup, since $\frac{1}{|E|}\hat{E}$ is a central idempotent.) But $g^2 e$ would not lie in FEe if g had infinite order. Thus, $g \in T$ and, in fact, $\alpha e = \lambda g \in FTe$. That is, $\alpha \in \mathscr{U}^+(FT)$. But T is abelian or a Hamiltonian 2-group, and either way, by Lemma 2.1.1, we see that the symmetric elements commute. We are done. □

Remark 2.5.7. It follows from the proof of the sufficiency above that if we replace the fourth condition with the assumption that every idempotent in FT is symmetric, then the symmetric units of FG commute.

2.6 The General Case for Nontorsion Groups

Let us now discuss group rings that are not semiprime. We let F be an infinite field of characteristic $p > 2$ (as the characteristic zero case is done). Once again, G is a group containing an element of infinite order, T is the set of torsion elements, P is the set of p-elements and Q is the set of p'-elements in T. As in the previous chapter, we will handle the case in which $N(FG)$ is nilpotent first.

Proposition 2.6.1. *Let F be an infinite field of characteristic $p > 2$ and G a group containing an element of infinite order. Suppose that $N(FG)$ is nilpotent. Then $\mathscr{U}^+(FG)$ satisfies a group identity if and only if P is a finite (normal) subgroup of G and $\mathscr{U}^+(F(G/P))$ satisfies a group identity.*

Proof. Suppose that $\mathscr{U}^+(FG)$ satisfies a group identity. By Proposition 1.2.22, $\phi_p(G)$ is finite. Thus, by Proposition 2.3.8, the p-elements of $\phi_p(G)$ form a group, so $\phi_p(G)$ is a finite p-group. Therefore, by Lemma 2.3.6, $\mathscr{U}^+(F(G/\phi_p(G)))$ satisfies a group identity. But by Proposition 1.2.9, $F(G/\phi_p(G))$ is semiprime. It follows from Theorem 2.5.6 that $G/\phi_p(G)$ has no p-elements. Thus, $P = \phi_p(G)$, and the necessity is proved. The sufficiency follows from Lemmas 2.3.5 and 1.1.1. □

Thus, in order to deal with groups containing only finitely many p-elements, it remains to show that the p-elements form a subgroup when $N(FG)$ is not nilpotent. We begin with

Proposition 2.6.2. *Let F be an infinite field of characteristic $p > 2$ and G a group. If $N(FG)$ is not nilpotent and $\mathscr{U}^+(FG)$ satisfies a group identity, then FG satisfies a polynomial identity.*

Proof. If G is torsion, then Proposition 2.4.3 does the job, so assume that G is not torsion. By Lemma 1.4.4, $T \cap \phi(G)$ is a locally finite group. Thus, by Proposition 2.3.8, its p-elements form a subgroup; hence, $\phi_p(G)$ is a p-group. Suppose it is finite. By Proposition 1.2.9, $F(G/\phi_p(G))$ is semiprime. Thus, $FG/\Delta(G, \phi_p(G))$ has no nilpotent ideals, hence $N(FG) \subseteq \Delta(G, \phi_p(G))$. By Lemma 1.1.1, $\Delta(G, \phi_p(G))$ is nilpotent, so $N(FG)$ is nilpotent, contrary to our assumption. Thus, $\phi_p(G) = \phi(G) \cap P$ is an infinite group.

By Lemma 2.2.6, $N(FG)$ satisfies a polynomial identity. As $N(FG)$ is not nilpotent, Lemma 1.2.16 tells us that FG satisfies a nondegenerate multilinear GPI. Thus, by Proposition 1.2.15, $(G : \phi(G)) < \infty$ and $|(\phi(G))'| < \infty$. Hence, by Proposition 1.1.4, it suffices to show that $F\phi(G)$ satisfies a polynomial identity. By Proposition 2.6.1, $N(F\phi(G))$ is not nilpotent. Thus, replacing G with $\phi(G)$, we assume that G is an FC-group. Then we know that G' is finite, so we can factor out $G' \cap P$ and assume that G' is a finite p'-group. Now, $(G, P) \leq G' \cap P = 1$. Thus, P is central. The last part of the proof of Proposition 1.5.9 finishes the argument. $\square$

As promised, we now have this proposition.

Proposition 2.6.3. *Let F be an infinite field of characteristic $p > 2$ and let G be a group. If $\mathscr{U}^+(FG)$ satisfies a group identity, then P is a (normal) subgroup of G and $\Delta(P)$ is locally nilpotent.*

Proof. If we can show that P is a subgroup, then we are done, since it is locally finite by Proposition 2.4.3, and therefore $\Delta(P)$ is locally nilpotent, by Lemma 1.1.1. If G is torsion, then Proposition 2.4.3 gives us the result. So, assume that G is nontorsion. If $N(FG)$ is nilpotent, then Proposition 2.6.1 does the job. Thus, we assume that $N(FG)$ is not nilpotent. By Proposition 2.6.2, FG satisfies a polynomial identity. Take any $g_1, \ldots, g_n \in P$ and let H be the subgroup they generate. Then FH is a finitely generated algebra satisfying a polynomial identity. By Proposition 1.5.12, $J(FH)$ is nilpotent. Now the Jacobson radical contains every nil ideal, so $N(FH)$ is nilpotent. By Propositions 2.4.3 and 2.6.1, H is a p-group. Thus, P is a group, and we are done. $\square$

As we have seen, if P is finite, there is nothing more to do. The second part of the main result of [96] is

Theorem 2.6.4. *Let F be an infinite field of characteristic $p > 2$ and G a nontorsion group. Suppose that G contains finitely many p-elements. If $\mathscr{U}^+(FG)$ satisfies a group identity, then*

1. the p-elements of G form a (finite normal) subgroup P of G;
2. the torsion elements of G/P form an abelian group, T/P;
3. every idempotent of F(T/P) is central in F(G/P); and
4. if F(T/P) has a nonsymmetric idempotent, then G/T satisfies a group identity.

Conversely, if G satisfies the four conditions above, and G/T is a u.p. group, then $\mathcal{U}^+(FG)$ satisfies a group identity.

What happens if P is infinite? In fact, if P has bounded exponent, then we are done as well. By Proposition 2.6.1, $N(FG)$ is not nilpotent. Hence, by Proposition 2.6.2, FG satisfies a polynomial identity (and therefore has a p-abelian normal subgroup of finite index, by Proposition 1.1.4). Thus, by Lemma 1.3.14, $\Delta(G,P)$ is nil of bounded exponent. By Lemma 2.3.5, $\mathcal{U}^+(FG)$ satisfies a group identity if and only if $\mathcal{U}^+(F(G/P))$ satisfies a group identity. As $F(G/P)$ is semiprime, Theorem 2.5.6 gives us the third part of the result.

Theorem 2.6.5. *Let F be an infinite field of characteristic $p > 2$ and G a nontorsion group. Suppose that G contains infinitely many p-elements and that the p-elements have bounded exponent. If $\mathcal{U}^+(FG)$ satisfies a group identity, then*

1. the p-elements of G form a (normal) subgroup P of bounded exponent;
2. the torsion elements of G/P form an abelian group, T/P;
3. every idempotent of F(T/P) is central in F(G/P);
4. if F(T/P) has a nonsymmetric idempotent, then G/T satisfies a group identity; and
5. G has a p-abelian normal subgroup of finite index.

Conversely, if G satisfies the five conditions above, and G/T is a u.p. group, then $\mathcal{U}^+(FG)$ satisfies a group identity.

Now we can consider groups in which P has unbounded exponent. We need to borrow the following general result about *-polynomial identities on prime algebras. The proof can be found in Theorems 2.4.13 and 3.1.62 of [91]. If $f(x_1, x_1^*, \ldots, x_n, x_n^*)$ is a *-polynomial in $F\{x_1, x_1^*, \ldots, x_n, x_n^*\}$, then we can construct a polynomial $g(x_1, \ldots, x_{2n}) \in F\{x_1, \ldots, x_{2n}\}$ by replacing each x_i^* with x_{n+i}. If f is a *-polynomial identity for an F-algebra R, then we say that f is special if g is a polynomial identity for R. (For example, let $R = M_2(F)$ under the transpose involution. Then R satisfies $[x_1 - x_1^*, x_2 - x_2^*]$, but surely not $[x_1 - x_3, x_2 - x_4]$. Thus, $[x_1 - x_1^*, x_2 - x_2^*]$ is not special for $M_2(F)$.)

Proposition 2.6.6. *Let F be an infinite field and R a prime F-algebra having an involution * fixing F elementwise. Suppose that R satisfies a polynomial identity. Then either*

1. R satisfies precisely the same *-polynomial identities as $M_n(F')$, under either the transpose or symplectic involution, for some positive integer n and some extension field F' of F; or
2. every *-polynomial identity of R is special.

Let us now refine Lemma 2.2.6.

Lemma 2.6.7. *Let F be an infinite field of characteristic different from 2 and let R be an F-algebra with an involution $*$ that fixes F elementwise. If I is a $*$-invariant nil ideal of R, let S be the F-subalgebra of R generated by 1 and I. Suppose that $\mathcal{U}^+(R)$ satisfies a group identity. Then S satisfies $*$-polynomial identities $f(x_1, x_1^*, \ldots, x_4, x_4^*)$ and $g(x_1, x_1^*, \ldots, x_4, x_4^*)$ such that*

1. *f is not a $*$-polynomial identity for $M_n(F)$ under the transpose involution, for any $n \geq 2$;*
2. *f is not a $*$-polynomial identity for $M_{2n}(F)$ under the symplectic involution for any $n \geq 2$; and*
3. *the sum of the monomials of g that do not involve any x_i^* is not a polynomial identity for $M_n(F)$ for any $n \geq 2$.*

Proof. Let $w(x, y) = 1$ be a group identity for $\mathcal{U}^+(R)$, as in Lemma 2.1.3, and let $F\{x_1, x_1^*, x_2, x_2^*\}[[z]]$ be the ring of formal power series over the free algebra with involution. We write

$$w((1+x_1 z)(1+x_1^* z), (1+x_2 z)(1+x_2^* z)) - 1 = \sum_{i \geq 0} f_i(x_1, x_1^*, x_2, x_2^*) z^i.$$

Now, if $a_1, a_2 \in I$, then for any $\lambda \in F$, we have $(1 + \lambda a_j)(1 + \lambda a_j^*) \in \mathcal{U}^+(R)$, for $j = 1, 2$. Thus,

$$0 = w((1+\lambda a_1)(1+\lambda a_1^*), (1+\lambda a_2)(1+\lambda a_2^*)) - 1 = \sum_{i \geq 0} f_i(a_1, a_1^*, a_2, a_2^*) \lambda^i.$$

Fixing $a_1, a_2 \in I$, and noting that if $a_1^k = 0$, then $(1 + \lambda a_1)^{-1} = 1 - \lambda a_1 + \lambda^2 a_1^2 - \cdots \pm \lambda^{k-1} a_1^{k-1}$, and similarly for the other terms, we see that there exists a positive integer j such that $f_i(a_1, a_1^*, a_2, a_2^*) = 0$ for all $i > j$. Thus,

$$\sum_{i=0}^{j} f_i(a_1, a_1^*, a_2, a_2^*) \lambda^i = 0.$$

As there are infinitely many choices for λ, a Vandermonde determinant argument (as in Lemma 1.2.4) tells us that each $f_i(a_1, a_1^*, a_2, a_2^*) = 0$, $0 \leq i \leq j$ (and obviously for all $i > j$). That is, I satisfies each f_i.

By definition of S, we have $[S, S] \subseteq I$. Thus, letting

$$g_i(x_1, x_1^*, \ldots, x_4, x_4^*) = f_i([x_1, x_3], [x_1, x_3]^*, [x_2, x_4], [x_2, x_4]^*)$$

for each i, we see that S satisfies each g_i. We will demonstrate that at least one g_i is not the zero polynomial by showing that it is not satisfied by a matrix ring.

Take any matrix ring $M_n(F)$, $n \geq 2$, with any involution. Suppose A is a square-zero matrix. Then noting that $(1 + Az)^s = 1 + sAz$ for all integers s and similarly for $1 + A^* z$, we have

$$\sum_{i \geq 0} f_i(A, A^*, A^*, A)z^i = w((1 + Az)(1 + A^*z), (1 + A^*z)(1 + Az)) - 1.$$

In this product we have no more than two consecutive identical terms, and no term is adjacent to its inverse. Replacing $(1 + Az)^s$ with $1 + sAz$, $s \in \{\pm 1, \pm 2\}$, we see that the highest term appearing is

$$\pm 2^u A^v (A^*A)^c (A^*)^d z^m = f_m(A, A^*, A^*, A)z^m,$$

with $u, c \geq 0$, $v, d \in \{0, 1\}$.

Suppose first of all that $*$ is the transpose involution. Then let $A = E_{12}$. Notice that

$$f_m(E_{12}, E_{21}, E_{21}, E_{12}) = \pm 2^u E_{12}^v (E_{21}E_{12})^c E_{21}^d = \pm 2^u E_{12}^v E_{22}^c E_{21}^d \neq 0.$$

If $M_n(F)$ satisfies g_m, then

$$0 = g_m(E_{11}, E_{11}, E_{22}, E_{22}, E_{12}, E_{21}, E_{21}, E_{12}) = f_m(E_{12}, E_{21}, E_{21}, E_{12}),$$

and we have a contradiction.

Next, suppose that $n = 2h$, $h \geq 2$, and $*$ is the symplectic involution. Let $A = E_{12} + E_{h+1,h+2}$. Notice that

$$f_m(E_{12} + E_{h+1,h+2}, E_{h+2,h+1} + E_{21}, E_{h+2,h+1} + E_{21}, E_{12} + E_{h+1,h+2})$$
$$= \pm 2^u (E_{12} + E_{h+1,h+2})^v (E_{22} + E_{h+2,h+2})^c (E_{h+2,h+1} + E_{21})^d.$$

As $E_{22} + E_{h+2,h+2}$ is an idempotent, this is not zero. If $M_n(F)$ satisfies g_m, then

$$0 = g_m(E_{11} + E_{h+1,h+1}, E_{h+1,h+1} + E_{11}, E_{h+2,h+2} + E_{22}, E_{22} + E_{h+2,h+2},$$
$$E_{12} + E_{h+1,h+2}, E_{h+2,h+1} + E_{21}, E_{h+2,h+1} + E_{21}, E_{12} + E_{h+1,h+2})$$
$$= f_m(E_{12} + E_{h+1,h+2}, E_{h+2,h+1} + E_{21}, E_{h+2,h+1} + E_{21}, E_{12} + E_{h+1,h+2}),$$

and again, we have a contradiction.

Finally, we claim that there exists an i such that if we let $\tilde{f}_i(x_1, x_2)$ be the sum of the monomials in f_i containing no x_r^* terms, then $\tilde{f}_i(E_{12}, E_{21}) \neq 0$ in $M_n(F)$, $n \geq 2$. If $A, B \in M_n(F)$ are square-zero, then calculating

$$\sum_{i \geq 0} f_i(A, A^*, B, B^*)z^i = w((1 + Az)(1 + A^*z), (1 + Bz)(1 + B^*z)) - 1,$$

we see that we do not get identical consecutive terms in the product, nor will a term appear next to its inverse. Replacing $(1 + Az)^{-1}$ with $1 - Az$, and so forth, and dropping all of the monomials involving A^* and B^*, the highest remaining term will be something of the form $\tilde{f}_s(A, B) = \pm A^v (BA)^c B^d$, with $c \geq 0$ and $v, d \in \{0, 1\}$. Letting $A = E_{12}$, $B = E_{21}$, we see that this is not zero, establishing the claim. Defining g_s as above, we see that $\tilde{g}_s(x_1, x_2, x_3, x_4) = \tilde{f}_s([x_1, x_3], [x_2, x_4])$. Evidently S satisfies g_s, but

$$\tilde{g}_s(E_{12}, E_{21}, E_{22}, E_{11}) = \tilde{f}_s(E_{12}, E_{21}) \neq 0.$$

We are done.   □

But we can be more specific. Note that every primitive ring is prime. Indeed, let R be primitive, and let I_1 and I_2 be nonzero ideals of R. If M is a faithful irreducible left R-module, then I_2M is a submodule of M. Since M is faithful, $I_2M \neq 0$, and since M is irreducible, this implies that $I_2M = M$. Similarly, $0 \neq M = I_1M = I_1I_2M$. Thus, $I_1I_2 \neq 0$, and R is prime.

Lemma 2.6.8. *Let F, R, I and S be as in the preceding lemma, with char $F = p > 2$. If $\mathscr{U}^+(R)$ satisfies a group identity, then there exists a positive integer k such that S satisfies the $*$-polynomial identities $[x_1 + x_1^*, x_2]^{p^k} = 0$ and $([x_1 + x_1^*, x_2]x_3)^{p^k} = 0$.*

Proof. Take f and g as in the preceding lemma. Let $F\{x_1, x_1^*, x_2, x_2^*, x_3, x_3^*\}$ be the free algebra with involution and let K be the ideal of this algebra generated by $f(a_1, a_1^*, \ldots, a_4, a_4^*)$ and $g(a_1, a_1^*, \ldots, a_4, a_4^*)$, together with their images under $*$, for all $a_i \in F\{x_1, x_1^*, x_2, x_2^*, x_3, x_3^*\}$. Let $W = R/K$. Then W has the induced involution and satisfies the $*$-polynomials f and g. Therefore, by Proposition 2.1.2, it satisfies a polynomial identity.

Let B be a primitive ideal of W; that is, let W/B be a primitive ring. Suppose that B is invariant under $*$. Then W/B has the induced involution and satisfies f. As we noted above, a primitive ring is prime. Thus, we examine the two possibilities allowed by Proposition 2.6.6. If $n \geq 2$, then f is not satisfied by $M_n(F')$ under the transpose involution nor by $M_{2n}(F')$ under the symplectic involution. Thus, if the first case of Proposition 2.6.6 occurs, then $n = 1$, hence we have a field or $M_2(F')$ under the symplectic involution. In the latter case, we note that the symmetric elements are simply scalar multiples of the identity matrix. Thus, W/B satisfies $[x_1 + x_1^*, x_2] = 0$. The alternative is that W/B satisfies the polynomial $h(x_1, \ldots, x_8)$ obtained by replacing x_i^* with x_{4+i} in f, $1 \leq i \leq 4$. But by Proposition 1.5.11, W/B satisfies the same polynomial identities as some $M_n(F')$. If $n \geq 2$, then we know that $M_n(F)$ under the transpose involution does not satisfy f, so take $A_1, \ldots, A_4 \in M_n(F')$ with $f(A_1, A_1^*, \ldots, A_4, A_4^*) \neq 0$. Then, substituting $x_i = A_i$ and $x_{4+i} = A_i^*$, $1 \leq i \leq 4$, we see that $M_n(F')$ does not satisfy h. It follows that $n = 1$ and W/B is commutative. Thus, in any case, W/B satisfies $[x_1 + x_1^*, x_2] = 0$.

Suppose, on the other hand, that $B^* \not\subseteq B$. Then $(B + B^*)/B$ is a nonzero ideal of W/B. But W/B is primitive and satisfies a polynomial identity. Thus, by Proposition 1.5.11, W/B is simple, so $W = B + B^*$. We can see that a typical element is $(b + b^*) + B = b^* + B$, $b \in B$. But $(b^*)^* \equiv 0 \pmod{B}$ for all $b \in B$. Thus, for any $b_1, \ldots, b_4 \in B^*$, we have

$$0 = g(b_1, b_1^*, \ldots, b_4, b_4^*) \equiv g(b_1, 0, \ldots, b_4, 0) \pmod{B}.$$

That is, letting $\tilde{g}$ be the sum of the monomials in g not involving any x_i^*, we have $\tilde{g}(b_1, b_2, b_3, b_4) \equiv 0 \pmod{B}$. Putting this another way, W/B satisfies $\tilde{g}$. By Proposition 1.5.11, W/B satisfies the same polynomial identities as some $M_n(F')$. But Lemma 2.6.7 tells us that $n = 1$. Thus, W/B is commutative. In all cases, then,

$[w_1 + w_1^*, w_2] \in B$ (and, therefore, $[w_1 + w_1^*, w_2]w_3 \in B$) for all $w_i \in W$ and all primitive ideals B of W.

The intersection of the primitive ideals is $J(W)$. Since W is a finitely generated F-algebra, Proposition 1.5.12 tells us that $J(W)$ is nilpotent. Thus, W satisfies the required $*$-polynomial identities. But W is the relatively free algebra with involution of the variety determined by f and g. Therefore, for any $s_1, s_2, s_3 \in S$, we see that $[s_1 + s_1^*, s_2]^{p^k} = 0$ and $([s_1 + s_1^*, s_2]s_3)^{p^k} = 0$, as required. □

Lemma 2.6.9. *Let F be an infinite field of characteristic $p > 2$ and G a nontorsion group with P infinite. Suppose that $\mathscr{U}^+(FG)$ satisfies a group identity and G has an abelian normal subgroup A of finite index. Then $\Delta(G, A \cap P)$ is nil and satisfies the $*$-polynomial identities $[x_1 + x_1^*, x_2]^{p^k} = 0$ and $([x_1 + x_1^*, x_2]x_3)^{p^k} = 0$ for some positive integer k.*

Proof. By Proposition 2.6.1, $N(FG)$ is not nilpotent, since P is infinite. Furthermore, Proposition 2.6.3 tells us that $\Delta(P)$ is locally nilpotent. The first part of the proof of Lemma 1.5.14 shows us that $\Delta(G, A \cap P)$ is locally nilpotent, hence nil. Thus, by the previous lemma, $\Delta(G, A \cap P)$ satisfies the required $*$-polynomial identities. □

We must show that G' has to be a p-group of bounded exponent. For now, we have

Lemma 2.6.10. *Let F be an infinite field of characteristic $p > 2$ and G a nontorsion group with P infinite. Suppose that $\mathscr{U}^+(FG)$ satisfies a group identity and G has an abelian normal subgroup A of finite index. Then $(G, A \cap P)$ is an abelian p-group of bounded exponent.*

Proof. Since A and P are normal subgroups, $(G, A \cap P)$ is clearly an abelian p-group. Thus, we need only check that it has bounded exponent. By the previous lemma, $\Delta(G, A \cap P)$ satisfies $[x_1 + x_1^*, x_2]^{p^k} = 0$. Choose any $a \in A \cap P$ and any $g \in G$. If we can bound the order of (a, g), then we will be done, as $(G, A \cap P)$ is abelian. We have

$$0 = [(a-1)g + ((a-1)g)^*, (a-1)g]^{p^k}$$
$$= (a + a^{-1} - a^g - (a^{-1})^g)^{p^k}$$
$$= a^{p^k} + a^{-p^k} - (a^{p^k})^g - (a^{-p^k})^g,$$

since A is abelian. Thus, $a^{p^k} + a^{-p^k} = (a^{p^k})^g + (a^{-p^k})^g$. That is, $a^{p^k} = (a^{p^k})^g$ or $(a^{-p^k})^g$. In the former case, $(a, g)^{p^k} = a^{-p^k}(a^{p^k})^g = 1$, as desired. Let us consider the latter case.

If $b = a^{p^k}$, then we have $b^{-1}g = gb$, hence $bg^2 = g^2b$. Thus,

$[(b-1)g+((b-1)g)^*, g(b-1)]$

$\quad = (b-1)g^2(b-1)+g^{-1}(b^{-1}-1)g(b-1)-g(b-1)^2g-g(b-1)g^{-1}(b^{-1}-1)$

$\quad = (b-1)^2g^2+(b-1)^2-(b^{-1}-1)^2g^2-(b^{-1}-1)^2$

$\quad = ((b-1)^2-(b^{-1}-1)^2)(g^2+1)$

$\quad = b^{-2}(b+1)(b-1)^3(g^2+1).$

Of course, $b \in A \cap P$. Therefore, $[(b-1)g+((b-1)g)^*, g(b-1)]^{p^k} = 0$, and hence

$$b^{-2p^k}(b^{p^k}+1)(b^{p^k}-1)^3(g^{2p^k}+1) = 0.$$

Of course b^{-2p^k} is a unit. Also, b^{p^k} is a p-element. If its order is p^r, $r \geq 0$, then $(b^{p^k}+1)^{p^r} = 2$. Thus, $b^{p^k}+1$ is also a unit. It follows that $(b^{p^k}-1)^3(g^{2p^k}+1) = 0$ and multiplying on the left by $(b^{p^k}-1)^{p-3}$, we get

$$(b^{p^{k+1}}-1)(g^{2p^k}+1) = 0.$$

That is,

$$b^{p^{k+1}}+b^{p^{k+1}}g^{2p^k} = g^{2p^k}+1$$

and hence

$$a^{p^{2k+1}}+a^{p^{2k+1}}g^{2p^k} = g^{2p^k}+1.$$

If $a^{p^{2k+1}} = 1$, then also $(a,g)^{p^{2k+1}} = a^{-p^{2k+1}}(a^g)^{p^{2k+1}} = 1$, as desired. Otherwise, $a^{p^{2k+1}}g^{2p^k} = 1$ and, therefore, $a^{p^{2k+1}} = g^{2p^k}$. That is, $a^{2p^{2k+1}} = 1$. Since $a \in P$, we have $a^{p^{2k+1}} = 1$ and the same conclusion follows. In any case, $(a,g)^{p^{2k+1}} = 1$, and we are done. $\square$

We close the chapter with the final part of the result of Sehgal and Valenti [96].

Theorem 2.6.11. *Let F be an infinite field of characteristic $p > 2$ and G a nontorsion group. Suppose that the p-elements of G are of unbounded exponent. Then the following are equivalent:*

(i) $\mathscr{U}^+(FG)$ satisfies a group identity;

(ii) $\mathscr{U}(FG)$ satisfies a group identity;

(iii) G has a p-abelian normal subgroup A of finite index, and G' is a p-group of bounded exponent.

Proof. Clearly (ii) implies (i), and we see from Theorem 1.5.16 that (iii) implies (ii). Thus, let us assume that $\mathscr{U}^+(FG)$ satisfies a group identity and show that G has a p-abelian normal subgroup A of finite index and that G' is a p-group of bounded exponent. By Proposition 2.6.1, $N(FG)$ is not nilpotent. Therefore, by Proposition 2.6.2, FG satisfies a polynomial identity, and the existence of A follows from Proposition 1.1.4. Now, A' is a finite p-group, so by Lemma 2.3.6, $\mathscr{U}^+(F(G/A'))$ satisfies a group identity. Furthermore, the p-elements of G/A' have unbounded exponent, and if $(G/A')'$ is a p-group of bounded exponent, then so is G'. Thus, we factor out A' and assume that A is abelian. Then, by Lemma 2.6.9, $\Delta(G, A \cap P)$ is nil. It follows from Lemma 2.3.5 that $\mathscr{U}^+(F(G/(A \cap P)))$ satisfies a group identity. Also, by Lemma 2.6.10, $(G, A \cap P)$ is a p-group of bounded exponent. Thus, we factor out $(G, A \cap P)$ and assume that $A \cap P$ is central (and of unbounded exponent, since $(P : A \cap P) \le (G : A) < \infty$). In fact, since P has a central subgroup of finite p-power index, Proposition 1.3.7 says that P' is a finite p-group. Thus, let us factor out P' and assume that P is abelian.

Take any $g, h \in G$ and $a \in A \cap P$. Since a is central, we have

$$(a - a^{-1})(g - g^{-1}) \in (\Delta(G, A \cap P))^+.$$

Also,

$$(a - a^{-1})h^{-1}, (a - a^{-1})^{p-2}gh \in \Delta(G, A \cap P).$$

Thus, by Lemma 2.6.9,

$$([(a - a^{-1})(g - g^{-1}), (a - a^{-1})h^{-1}](a - a^{-1})^{p-2}gh)^{p^k} = 0.$$

Again, a is central, so we have

$$(a - a^{-1})^{p^{k+1}}([g - g^{-1}, h^{-1}]gh)^{p^k} = 0.$$

Multiplying through by $a^{p^{k+1}}$, we get

$$(a^{2p^{k+1}} - 1)([g - g^{-1}, h^{-1}]gh)^{p^k} = 0.$$

Recall that $A \cap P$ has unbounded exponent. Thus, we have infinitely many different $a^{2p^{k+1}}$; hence, by Lemma 1.5.8, $([g - g^{-1}, h^{-1}]gh)^{p^k} = 0$. Equivalently,

$$(gh^{-1}gh - (g, h) - h^{-1}g^2h + 1)^{p^k} = 0.$$

Since this holds for any $g \in G$, it also holds for ag. Thus,

$$(a^2gh^{-1}gh - (g, h) - a^2h^{-1}g^2h + 1)^{p^k} = 0.$$

By Lemma 1.3.16,

$$a^{2p^k}(gh^{-1}gh)^{p^k} - (g, h)^{p^k} - a^{2p^k}(h^{-1}g^2h)^{p^k} + 1 \in [FG, FG].$$

But elements of $[FG, FG]$ must have trace zero, so either $(g, h)^{p^k}$ or $a^{2p^k}(h^{-1}g^2h)^{p^k}$ is the identity. However, $A \cap P$ has unbounded exponent, so we can choose a in such a way that $a^{2p^k} \ne (h^{-1}g^2h)^{-p^k}$. Thus, $(g, h)^{p^k} = 1$ for all $g, h \in G$. But $G' \le P$, and P is abelian. Thus, G' is a p-group of bounded exponent, and the proof is complete. $\qquad\square$

Chapter 3
Lie Identities on Symmetric Elements

3.1 Introduction and Classical Results

We now pause in our discussion of units in group rings to consider Lie properties instead. This is an important digression. Indeed, we have already seen the connection between group identities on units (or symmetric units) and polynomial identities on the group ring. We will see in subsequent chapters that we can frequently reduce problems concerning specific group identities to problems concerning specific Lie identities.

Recall that on any ring R, we define the Lie product via $[x_1, x_2] = x_1 x_2 - x_2 x_1$. We can extend this recursively via

$$[x_1, \ldots, x_{n+1}] = [[x_1, \ldots, x_n], x_{n+1}].$$

Let S be a subset of R. We say that S is Lie nilpotent if there exists an $n \geq 2$ such that $[a_1, \ldots, a_n] = 0$ for all $a_i \in S$. For a positive integer n, we say that S is Lie n-Engel if

$$[a, \underbrace{b, \ldots, b}_{n \text{ times}}] = 0$$

for all $a, b \in S$. The set S is bounded Lie Engel if it is Lie n-Engel for some positive integer n.

We can define a different operation as follows. Let $[x_1, x_2]^o = [x_1, x_2]$ and for any positive integer n, let

$$[x_1, \ldots, x_{2^{n+1}}]^o = [[x_1, \ldots, x_{2^n}]^o, [x_{2^n+1}, \ldots, x_{2^{n+1}}]^o].$$

Then S is Lie solvable if there exists a positive integer n such that $[a_1, \ldots, a_{2^n}]^o = 0$ for all $a_i \in S$.

In our case, we are most interested in the set of symmetric elements $(FG)^+$ of a group ring FG. We will discuss the conditions under which $(FG)^+$ is Lie nilpotent, bounded Lie Engel or Lie solvable in this chapter. As these properties are

G.T. Lee, *Group Identities on Units and Symmetric Units of Group Rings*,
Algebra and Applications 12, DOI 10.1007/978-1-84996-504-0_3,
© Springer-Verlag London Limited 2010

*-polynomial identities for FG, we can use Proposition 2.1.2 to see that FG satisfies a polynomial identity. Few results concerning the Lie properties of $(FG)^+$ are known if char $F = 2$. We will include those results that are known. However, unless otherwise indicated, we shall assume that char $F = p \neq 2$.

The conditions under which FG has the above properties were determined in classical papers. Passi et al. discovered when FG is Lie solvable or Lie nilpotent in [81], and Sehgal determined when FG is bounded Lie Engel in [94]. The proofs of these results occupy Chapter V in [94], and we will not reproduce them here. The results are as follows.

Theorem 3.1.1. *Let F be a field of characteristic $p \geq 0$ and G a group. Then FG is Lie nilpotent if and only if G is nilpotent and p-abelian.*

Proof. See [94, Theorem V.4.4]. □

Theorem 3.1.2. *If char $F = 0$, then FG is bounded Lie Engel if and only if G is abelian. If char $F = p > 0$, then FG is bounded Lie Engel if and only if G is nilpotent and G has a p-abelian normal subgroup of finite p-power index.*

Proof. See [94, Theorem V.6.1]. □

Theorem 3.1.3. *If char $F \neq 2$, then FG is Lie solvable if and only if G is p-abelian. If char $F = 2$, then FG is Lie solvable if and only if G has a 2-abelian subgroup of index at most 2.*

Proof. See [94, Theorem V.3.1]. □

Work on the Lie nilpotence of $(FG)^+$ was begun by Giambruno and Sehgal in [36]. They showed that if G has no 2-elements and $(FG)^+$ is Lie nilpotent, then FG is Lie nilpotent. It is easy enough to see that this will not hold if G has 2-elements. Indeed, if G is a Hamiltonian 2-group, then by Lemma 2.1.1, the symmetric elements of FG commute. But Theorem 3.1.1 tells us that FG is not Lie nilpotent. In [64], the author showed that their result can be extended to groups not containing the quaternions and then classified the groups containing Q_8 such that $(FG)^+$ is Lie nilpotent.

Work by the author on the bounded Lie Engel property in [65] came a bit later, but we will present it first, since we can then use this result to cut down on the work required for Lie nilpotence.

Finally, we will present the results due to Lee et al. [71] and Lee and Spinelli [73] concerning Lie solvability of the symmetric elements. Here, a restriction upon the orders of the group elements will be imposed, as no fully general result is currently known.

Before we begin the main part of our work in the next section, let us record a few simple observations of which we will make use in the future without further mention.

Lemma 3.1.4. *Suppose char $F \neq 2$ and G is a group with H any subgroup and N a normal subgroup. If $(FG)^+$ is Lie nilpotent (resp., bounded Lie Engel, Lie solvable), then so are $(FH)^+$ and $(F(G/N))^+$.*

Proof. We see that $(FH)^+$ is a subset of $(FG)^+$, and thus inherits the required property. For $F(G/N)$, simply observe that the symmetric elements are linear combinations of terms of the form $gN + g^{-1}N$, with $g \in G$. That is, every element of $(F(G/N))^+$ is the homomorphic image of an element of $(FG)^+$ under the natural map $FG \to F(G/N)$. The result follows immediately. $\qquad\square$

Also, the characteristic zero case tends to be easy, as we can reduce to $\mathbb{Z}_p$ for any prime p.

Lemma 3.1.5. *Suppose char $F = 0$ and G is any group. If $(FG)^+$ is Lie nilpotent (resp., bounded Lie Engel, Lie solvable), then so is $(\mathbb{Z}_p G)^+$ for any prime p.*

Proof. Clearly $(\mathbb{Z}G)^+ \subseteq (FG)^+$, hence $(\mathbb{Z}G)^+$ is Lie nilpotent (resp., bounded Lie Engel, Lie solvable). But there is a natural homomorphism $\mathbb{Z}G \to \mathbb{Z}_p G$, and every symmetric element of $\mathbb{Z}_p G$ is the homomorphic image of a symmetric element of $\mathbb{Z}G$. The result follows immediately. $\qquad\square$

One more lemma will be of inestimable value.

Lemma 3.1.6. *Let R be a ring of prime characteristic p. Then for any $a, b \in R$ and any $n \geq 0$, we have*
$$[a, \underbrace{b, \ldots, b}_{p^n \text{ times}}] = [a, b^{p^n}].$$

Proof. Fix $b \in R$. Let $\theta_r : R \to R$ be defined via $\theta_r(c) = cb$ for all $c \in R$. Similarly, let $\theta_l : R \to R$ be defined via $\theta_l(c) = bc$ for all $c \in R$. Then we notice that $[c, b] = (\theta_r - \theta_l)(c)$ for all $c \in R$. Thus,
$$[a, \underbrace{b, \ldots, b}_{p^n \text{ times}}] = (\theta_r - \theta_l)^{p^n}(a).$$

But θ_r and θ_l are commuting functions, and R has characteristic p; hence
$$(\theta_r - \theta_l)^{p^n}(a) = (\theta_r^{p^n} - \theta_l^{p^n})(a) = ab^{p^n} - b^{p^n}a,$$

as required. $\qquad\square$

3.2 The Bounded Lie Engel Property

Let F be a field of characteristic $p \neq 2$ and G a group. We present the results of the author in [65] determining when $(FG)^+$ is bounded Lie Engel. The techniques used borrow freely from [36], [64] and [81] as well.

We begin with

Lemma 3.2.1. *Let F be a field of characteristic different from 2 and G a group. If $(FG)^+$ is bounded Lie Engel, then for every element $g \in G$ of order 2, we have $g \in \zeta(G)$.*

Proof. As we have seen, the characteristic zero case will follow from the prime characteristic case, so suppose $p > 2$ and let $(FG)^+$ be Lie p^n-Engel. Take $g \in G$ of order 2. If $h \in G$ has order 2 as well, then we get

$$0 = [h, \underbrace{g, \ldots, g}_{p^n \text{ times}}] = [h, g^{p^n}] = [h, g],$$

since p^n is odd. Thus, g and h commute.

Suppose, on the other hand, that h does not have order 2. Then

$$0 = [h + h^{-1}, \underbrace{g, \ldots, g}_{p^n \text{ times}}] = [h + h^{-1}, g^{p^n}] = [h + h^{-1}, g].$$

That is,

$$hg + h^{-1}g = gh + gh^{-1},$$

hence $hg = gh$ or gh^{-1}. In the latter case, $(hg)^2 = hg^2h^{-1} = 1$. Thus, as we have seen, hg commutes with g. But since $ghg = hg^2$, we have $gh = hg$ in any event. That is, g is central, as required. $\qquad\square$

Lemma 3.2.2. *Suppose char $F \neq 2$. If $g, h \in G$ satisfy $[g + g^{-1}, h + h^{-1}] = 0$, then $gh \in \{hg, hg^{-1}, h^{-1}g, h^{-1}g^{-1}\}$.*

Proof. We have

$$gh + g^{-1}h + gh^{-1} + g^{-1}h^{-1} = hg + h^{-1}g + hg^{-1} + h^{-1}g^{-1}.$$

As each side is a sum of group elements, the conclusion follows immediately unless char $F = 3$. In that case, we must consider the possibility that exactly three terms on the left-hand side agree. But if that is so, then either $gh = g^{-1}h$ or $gh^{-1} = g^{-1}h^{-1}$. Either way, $g = g^{-1}$, and it follows easily that if three of the terms on the left-hand side agree, then all four agree, giving us a contradiction. $\qquad\square$

Groups containing Q_8 will be handled separately from those not containing Q_8. First, let us consider groups not containing Q_8. A common situation that can then be avoided is described in the next lemma.

Lemma 3.2.3. *Suppose char $F \neq 2$ and G is a group generated by elements g and h, with $g^h = g^{-1}$. Let $(FG)^+$ be bounded Lie Engel. Then either $g^2 = 1$ and G is abelian, or $o(g) = 4$, $o(h) = 4n$ for some odd number n, and $\langle g, h^n \rangle \simeq Q_8$.*

Proof. We may assume that $g^2 \neq 1$. Notice that $g^i \in \zeta(G)$ if and only if $g^{2i} = 1$ and that $h^j \in \zeta(G)$ if and only if j is even.

As usual, the characteristic zero case will follow from the prime characteristic case, so let char $F = p > 2$ and suppose that $(FG)^+$ is Lie p^m-Engel. We have

$$0 = [gh + (gh)^{-1}, \underbrace{h + h^{-1}, \ldots, h + h^{-1}}_{p^m \text{ times}}] = [gh + (gh)^{-1}, h^{p^m} + h^{-p^m}].$$

By the previous lemma, we have four possible values for gh^{p^m+1}.

First, if $gh^{p^m+1} = h^{p^m}gh$, then $gh^{p^m} = h^{p^m}g$. That is, h^{p^m} is central. But p^m is odd, and we have a contradiction.

Second, if $gh^{p^m+1} = h^{p^m-1}g^{-1}$, then as h^{p^m+1} is central, we have $h^{p^m+1}g = h^{p^m-1}g^{-1}$. That is, $g^2 = h^{-2}$.

Third, suppose that $gh^{p^m+1} = h^{-p^m}gh$. Then $gh^{p^m} = h^{-p^m}g$. But $g^{h^{p^m}} = g^{-1}$, since p^m is odd. Thus, $h^{p^m}g^{-1} = h^{-p^m}g$, and $g^2 = h^{2p^m}$.

Finally, if $gh^{p^m+1} = h^{-p^m-1}g^{-1}$, then $(gh^{p^m+1})^2 = 1$. By Lemma 3.2.1, gh^{p^m+1} is central. But h^{p^m+1} is also central and, therefore, g is central, contrary to our assumption.

Thus, in any case, $g^2 = h^{2k}$ for some odd integer k. It follows that $g^2 \in \zeta(G)$ and therefore $g^4 = 1$. We are assuming that $g^2 \neq 1$, hence $o(g) = 4$. Thus, for some odd positive integer n, $o(h) = 4n$. That is, $o(g) = 4$, $g^2 = (h^n)^2$, and $g^{h^n} = g^{-1}$, hence $\langle g, h^n \rangle$ is the quaternion group. $\qquad \square$

Combining the last few lemmas, we obtain the following result.

Lemma 3.2.4. *Let char $F \neq 2$ and suppose that $(FG)^+$ is bounded Lie Engel. If G does not contain Q_8, then for any $g, h \in G$ satisfying $[g + g^{-1}, h + h^{-1}] = 0$, we must have $gh = hg$.*

Proof. Lemma 3.2.2 gives us four possibilities for gh. If $gh = hg$, there is nothing to say. If $gh = hg^{-1}$ or $h^{-1}g$, then the previous lemma applied to $\langle g, h \rangle$ tells us that $gh = hg$. Finally, if $gh = h^{-1}g^{-1}$, then $(gh)^2 = 1$; hence, by Lemma 3.2.1, $gh \in \zeta(G)$. Thus, $ghg = g^2h$, and $hg = gh$, as required. $\qquad \square$

We can now dispense with the characteristic zero case (for groups not containing the quaternions) and make a useful reduction for fields of prime characteristic.

Lemma 3.2.5. *Let char $F = p \neq 2$ and suppose that G is a group not containing Q_8 such that $(FG)^+$ is bounded Lie Engel. If $p = 0$, then G is abelian. If $p > 2$, then $G/\zeta(G)$ is a p-group of bounded exponent.*

Proof. Suppose $p > 2$ and let $(FG)^+$ be Lie p^n-Engel. Take any $g, h \in G$. We have

$$0 = [g + g^{-1}, \underbrace{h + h^{-1}, \ldots, h + h^{-1}}_{p^n \text{ times}}] = [g + g^{-1}, h^{p^n} + h^{-p^n}].$$

By the last lemma, h^{p^n} commutes with g. But g and h were arbitrary, so $G^{p^n} \subseteq \zeta(G)$.

If char $F = 0$, then by reducing to the characteristic 3 case, we get that $G/\zeta(G)$ is a 3-group, and by reducing to the characteristic 5 case, we see that $G/\zeta(G)$ is a 5-group. Thus, $G = \zeta(G)$. $\qquad \square$

In order to finish the case without quaternions, we need one group-theoretic result. If $H_1, H_2, \ldots$ are subgroups of G, then we write, as usual, (H_1, H_2) for the subgroup of G generated by all (h_1, h_2) with $h_i \in H_i$, and $(H_1, \ldots, H_k) = ((H_1, \ldots, H_{k-1}), H_k)$. For any subgroup H of G, we write $C_G(H)$ for the centralizer of H in G. We have

Lemma 3.2.6. *Let G be a group and A an abelian normal subgroup. Suppose, for some prime p, that $G/C_G(A)$ is a finite p-group and A is a p-group of bounded exponent. Then there exists a positive integer n such that*

$$(A,\underbrace{G,\ldots,G}_{n\ times}) = 1.$$

Proof. Let p^m be the exponent of A. Then we can make A into a $\mathbb{Z}_{p^m}G$-module in the following way: namely, if $\alpha = \sum_{g \in G} \alpha_g g \in \mathbb{Z}_{p^m}G$, then for any $a \in A$, we let

$$a^\alpha = \prod_{g \in G}(a^{\alpha_g})^g.$$

Let N be any normal subgroup of G contained in A. Then taking $a \in N$, $g \in G$, we have $(a,g) = a^{-1}a^g = a^{-1+g}$. But this lies in $N^{\Delta(G)}$; that is, the subgroup of N generated by all elements of the form a^δ, with $a \in N$, $\delta \in \Delta(G)$. Thus, $(A,G) \subseteq A^{\Delta(G)}$, $(A,G,G) \subseteq A^{(\Delta(G))^2}$, and so forth.

Now, $G/C_G(A)$ is a finite p-group. Thus, by Lemma 1.1.1, $\Delta(G/C_G(A))$ is nilpotent. Hence, there exists an n so that $(\Delta(G))^n \subseteq \Delta(G,C_G(A))$. Thus,

$$(A,\underbrace{G,\ldots,G}_{n\ times}) \subseteq A^{\Delta(G,C_G(A))}.$$

In particular, every element of $(A,G,\ldots,G)$ is a product of terms of the form $a^{(h-1)g}$, with $a \in A$, $h \in C_G(A)$ and $g \in G$. But

$$a^{(h-1)g} = (a^h)^g(a^{-1})^g = a^g(a^{-1})^g = 1.$$

We are done.   □

The following lemma is also vital.

Lemma 3.2.7. *Let char $F = p > 0$ and let G be a p-group of bounded exponent such that FG satisfies a nondegenerate multilinear GPI. Then G is nilpotent.*

Proof. Letting $A = \phi(G)$, Proposition 1.2.15 tells us that A is of finite index in G and A' is finite. In particular, $(G/A')/C_{G/A'}(A/A')$ is finite, and the previous lemma applies. We have

$$(A/A',G/A',\ldots,G/A') = 1,$$

hence

$$(A,G,\ldots,G) \le A'.$$

Now, A'/A'' is a finite subgroup of $\phi(G/A'')$ and therefore, $C_{G/A''}(A'/A'')$ has finite index in G/A''. Again applying the previous lemma, we get

$$(A'/A'',G/A'',\ldots,G/A'') = 1.$$

That is,

$$(A',G,\ldots,G) \le A'',$$

hence

$$(A,G,\ldots,G) \le A''.$$

Repeating this procedure, and noting that A' is nilpotent (being a finite p-group), we eventually get

$$(A,G,\ldots,G) = 1.$$

But G/A is a finite p-group, so $(G,\ldots,G) \le A$, and we conclude that $(G,\ldots,G) = 1$, as required.   □

The first part of the main result of Lee [65] is

Theorem 3.2.8. *Let F be a field of characteristic $p \ne 2$ and G a group not containing the quaternions. Then the following are equivalent:*

(i) $(FG)^+$ is bounded Lie Engel;
(ii) FG is bounded Lie Engel;
(iii) a. $p = 0$ and G is abelian, or
   b. $p > 2$, G is nilpotent, and G has a p-abelian normal subgroup of finite p-power index.

Proof. In view of Theorem 3.1.2, it remains only to show that (i) implies (iii). The characteristic zero case follows immediately from Lemma 3.2.5, so let $p > 2$ and suppose that $(FG)^+$ is bounded Lie Engel.

We claim, first of all, that G is nilpotent. It suffices to show that $G/\zeta(G)$ is nilpotent. But by Lemma 3.2.5, $G/\zeta(G)$ is a p-group of bounded exponent. Also, we know that FG satisfies a $*$-polynomial identity of the form

$$[x_1 + x_1^*, \underbrace{x_2 + x_2^*, \ldots, x_2 + x_2^*}_{n \text{ times}}] = 0,$$

and therefore, by Proposition 2.1.2, it satisfies a polynomial identity. Applying the previous lemma, we see that G is nilpotent.

It follows from Lemma 3.2.5 and Proposition 1.3.7 that G' is a p-group. Also, by Proposition 1.2.15, $(G : \phi(G)) < \infty$ and $|(\phi(G))'| < \infty$. But $\zeta(G) \le \phi(G)$, and we know that $G/\zeta(G)$ is a p-group. Thus, $G/\phi(G)$ and $(\phi(G))'$ are both finite p-groups, and we are done.   □

We also mention that in the author's thesis [66], the characteristic 2 case was handled when G has no 2-elements. As the proof is quite short and not published elsewhere, let us include it here.

We need the following lemma from Giambruno and Sehgal [36], which will be important in dealing with the other Lie properties. It does not depend upon the characteristic of the field. If $f(x_1, x_1^*, \ldots, x_n, x_n^*)$ is a $*$-polynomial identity, then we say that it is linear in x_i if, in each monomial of f, exactly one of $\{x_i, x_i^*\}$ appears, and it appears only once in each monomial. For instance, $x_1 x_2 + x_1^* x_2 x_2^*$ is linear in x_1, but not x_2.

Lemma 3.2.9. *Let F be a field and G a group such that the set $\{g^2 : g \in \zeta(G)\}$ is infinite. Suppose that FG satisfies a $*$-polynomial identity $f(x_1, x_1^*, \ldots, x_n, x_n^*)$ that is linear in x_1. Then FG also satisfies the $*$-polynomial obtained by deleting all of the monomials in which x_1^* appears.*

Proof. Let us write $f = f_1 + f_2$, where x_1 appears in each monomial of f_1 and x_1^* appears in each monomial of f_2. Take $\alpha_1, \ldots, \alpha_n \in FG$ and $z \in \zeta(G)$. Then

$$
\begin{aligned}
0 &= f(z\alpha_1, (z\alpha_1)^*, \alpha_2, \alpha_2^*, \ldots, \alpha_n, \alpha_n^*) \\
&= f_1(z\alpha_1, z^{-1}\alpha_1^*, \alpha_2, \alpha_2^*, \ldots, \alpha_n, \alpha_n^*) + f_2(z\alpha_1, z^{-1}\alpha_1^*, \alpha_2, \alpha_2^*, \ldots, \alpha_n, \alpha_n^*) \\
&= z f_1(\alpha_1, \alpha_1^*, \ldots, \alpha_n, \alpha_n^*) + z^{-1} f_2(\alpha_1, \alpha_1^*, \ldots, \alpha_n, \alpha_n^*).
\end{aligned}
$$

But also,

$$
0 = z^{-1}(f_1(\alpha_1, \alpha_1^*, \ldots, \alpha_n, \alpha_n^*) + f_2(\alpha_1, \alpha_1^*, \ldots, \alpha_n, \alpha_n^*)).
$$

Subtracting, we get

$$
(z - z^{-1}) f_1(\alpha_1, \alpha_1^*, \ldots, \alpha_n, \alpha_n^*) = 0,
$$

hence

$$
(z^2 - 1) f_1(\alpha_1, \alpha_1^*, \ldots, \alpha_n, \alpha_n^*) = 0.
$$

But there are infinitely many such z^2. Thus, Lemma 1.5.8 gives us our result.   □

We can now prove

Lemma 3.2.10. *Let F be a field of characteristic 2 and let $G = \langle g, h \rangle$ be a group without 2-elements. If $g \neq 1$ and $g^h = g^{-1}$, then $(FG)^+$ is not bounded Lie Engel.*

Proof. Since G has no 2-elements, $g^2 \neq 1$. Thus, h^2 is central, but h is not. Since h cannot have even order, it must have infinite order. Thus, h^2 is an element of infinite order in $\zeta(G)$. Suppose the result does not hold, and choose n so that $(FG)^+$ is Lie 2^n-Engel. Then, by the preceding lemma, FG satisfies the $*$-polynomial identity

$$
0 = [x, \underbrace{y + y^*, \ldots, y + y^*}_{2^n \text{ times}}] = [x, (y + y^*)^{2^n}].
$$

Thus, we see that

$$
[h, (gh^2)^{2^n} + (gh^2)^{-2^n}] = 0.
$$

Expanding, we get

$$
h^{2^{n+1}+1} g^{2^n} + h^{1-2^{n+1}} g^{-2^n} + g^{2^n} h^{2^{n+1}+1} + g^{-2^n} h^{1-2^{n+1}} = 0.
$$

There are three possibilities to consider.

If $h^{2^{n+1}+1} g^{2^n} = h^{1-2^{n+1}} g^{-2^n}$, then $g^{2^{n+1}} \in \langle h \rangle$, hence $g^{2^{n+1}} \in \zeta(G)$. But $(g^{2^{n+1}})^h = g^{-2^{n+1}}$, hence $g^{2^{n+2}} = 1$. As G has no 2-elements, this gives a contradiction.

If $h^{2^{n+1}+1}g^{2^n} = g^{2^n}h^{2^{n+1}+1} = h^{2^{n+1}}g^{2^n}h$, then $hg^{2^n} = g^{2^n}h$, and g^{2^n} is central. Once again, this implies that $g^{2^{n+1}} = 1$, and this is impossible.

Finally, suppose that $h^{2^{n+1}+1}g^{2^n} = g^{-2^n}h^{1-2^{n+1}}$. Since $(g^{2^n})^h = g^{-2^n}$, we get $g^{-2^n}h^{2^{n+1}+1} = g^{-2^n}h^{1-2^{n+1}}$. That is, h is torsion, which gives us a contradiction. We are done.   □

Lemma 3.2.11. *Let* char $F = 2$ *and let* $G = \langle g,h \rangle$ *be a group containing no 2-elements, such that* $(FG)^+$ *is bounded Lie Engel. If* $[g+g^{-1}, h+h^{-1}] = 0$, *then* $gh = hg$.

Proof. Expanding the equation, we get

$$gh + g^{-1}h + gh^{-1} + g^{-1}h^{-1} + hg + hg^{-1} + h^{-1}g + h^{-1}g^{-1} = 0.$$

Thus, gh is equal to one of the other group elements.

If $gh = g^{-1}h$, then $g^2 = 1$, hence $g = 1$, and there is nothing to do. Similarly if $gh = gh^{-1}$. If $gh = h^{-1}g^{-1}$, then $(gh)^2 = 1$, hence $gh = 1$, and again, there is nothing to do. If $gh = hg^{-1}$ or $h^{-1}g$, then the previous lemma applies. Of course, if $gh = hg$, then we are done.

Thus, we may assume that $gh = g^{-1}h^{-1}$, hence $g^2 = h^{-2} \in \zeta(G)$. If $o(g) < \infty$, then since g must have odd order, we have $g \in \zeta(G)$, as desired. So, suppose that $o(g) = \infty$. Then

$$0 = [g + g^{-1}, h + h^{-1}] = (1 + g^{-2})(1 + h^{-2})[g,h].$$

Thus, for any positive integer m,

$$(1 + g^{-2^m})(1 + h^{-2})[g,h] = (1 + g^{-2} + g^{-4} + \cdots + g^{2-2^m})(1 + g^{-2})(1 + h^{-2})[g,h]$$
$$= 0.$$

As g has infinite order, Lemma 1.5.8 says that $(1 + h^{-2})[g,h] = 0$. As h also has infinite order, we can apply the same argument to deduce that $[g,h] = 0$, and we are done.   □

We can now prove the characteristic two result as before.

Theorem 3.2.12. *Let* F *be a field of characteristic 2 and* G *a group without 2-elements. Then* $(FG)^+$ *is bounded Lie Engel if and only if* G *is abelian.*

Proof. Suppose that $(FG)^+$ is bounded Lie Engel. In view of the last lemma, we can use the proof of Lemma 3.2.5 to show that $G/\zeta(G)$ is a 2-group of bounded exponent. Of course, FG satisfies a $*$-polynomial identity, hence, by Proposition 2.1.2, a polynomial identity. Therefore, $F(G/\zeta(G))$ satisfies a polynomial identity. Thus, by Lemma 3.2.7, G is nilpotent, and by Proposition 1.3.7, G' is a 2-group. As G has no 2-elements, it must be abelian. The converse is obvious.   □

Remark 3.2.13. In [65], the author also showed by similar means that if char $F \neq 2$ and G has no 2-elements, then the skew elements of FG are bounded Lie Engel if and only if FG is bounded Lie Engel. Clearly, we cannot extend this even to groups without quaternions. Indeed, let G be any dihedral group. Then the skew elements are easily seen to commute, but by Theorem 3.1.2, FG is not bounded Lie Engel. If G is a finite group with FG semisimple, then the author determined the conditions under which $(FG)^-$ is bounded Lie Engel in [68]. For other groups containing 2-elements, the solution is not currently known.

Let us return to fields of characteristic different from 2 and consider groups containing the quaternions. By Theorem 3.1.2, FG cannot be bounded Lie Engel. We assume that $(FG)^+$ is bounded Lie Engel and begin to narrow the possibilities.

Lemma 3.2.14. *Let char $F = p > 2$, and let $G = Q_8 \times \langle c \rangle$. If $(FG)^+$ is Lie p^m-Engel, then the order of c divides $2p^m$.*

Proof. Let $Q_8 = \langle g, h \rangle$. Then

$$0 = [gc + (gc)^{-1}, \underbrace{hc + (hc)^{-1}, \ldots, hc + (hc)^{-1}}_{p^m \text{ times}}] = [gc + (gc)^{-1}, (hc)^{p^m} + (hc)^{-p^m}].$$

In view of Lemma 3.2.2, there are four possibilities for $gh^{p^m} c^{p^m+1}$.

First, if $gh^{p^m} c^{p^m+1} = h^{p^m} g c^{p^m+1}$, then g commutes with h^{p^m}. But p^m is odd, so this is not the case.

Second, if $gh^{p^m} c^{p^m+1} = h^{p^m} g^{-1} c^{p^m-1}$, then $c^2 \in \langle g, h \rangle \cap \langle c \rangle = 1$. Thus, the assertion holds.

Third, if $gh^{p^m} c^{p^m+1} = h^{-p^m} g c^{1-p^m}$, then $c^{2p^m} = 1$, as required.

Finally, if $gh^{p^m} c^{p^m+1} = h^{-p^m} g^{-1} c^{-p^m-1}$, then $(gh^{p^m})^2 = 1$. But this is impossible, so we are done.   □

Lemma 3.2.15. *Let G be a group containing $Q_8 = \langle g, h \rangle$, and let char $F = p > 2$. Suppose $(FG)^+$ is Lie p^n-Engel. If $a \in G$ does not centralize $\langle g, h \rangle$, then $o(a) = 4p^m$ for some $0 \leq m \leq n$, and each of $\langle g, a^{p^m} \rangle$ and $\langle h, a^{p^m} \rangle$ is abelian or isomorphic to Q_8.*

Proof. As usual, we get

$$0 = [a + a^{-1}, g^{p^n} + g^{-p^n}] = [a + a^{-1}, g + g^{-1}],$$

since $o(g) = 4$. Then Lemma 3.2.2 gives us four possibilities for ag.

First, we could have $ag = ga$.

Second, suppose that $ag = ga^{-1}$. Then $a^g = a^{-1}$. By Lemma 3.2.3, either $a^2 = 1$ (hence $ag = ga$) or $o(a) = 4$ and $\langle a, g \rangle \simeq Q_8$. This is our assertion with $m = 0$.

Third, suppose that $ag = g^{-1}a$. Then $g^a = g^{-1}$. As $o(g) > 2$, Lemma 3.2.3 states that $o(a) = 4k$, where k is odd, and $\langle a^k, g \rangle \simeq Q_8$. Evidently, a^4 centralizes $\langle a^k, g \rangle$. Since $o(a^4) = k$, the product $\langle g, a^k \rangle \times \langle a^4 \rangle$ must be direct. By the previous lemma, $k = p^m$ for some $m \leq n$.

Finally, if $ag = g^{-1}a^{-1}$, then $(ag)^2 = 1$ and ag is central, by Lemma 3.2.1. Thus, $gag = ag^2$, and $ga = ag$.

Repeating this procedure with h in place of g, we obtain our conclusion. $\square$

Let us now restrict the 2-elements of G.

Lemma 3.2.16. *Suppose char $F \neq 2$ and $Q_8 \leq G$. If $(FG)^+$ is bounded Lie Engel, then the 2-elements of G form a (normal) subgroup which is a Hamiltonian 2-group.*

Proof. As the characteristic zero case follows from the characteristic p case, let char $F = p > 2$. Let $a, b \in G$ be 2-elements. Choose n in such a way that $(FG)^+$ is Lie p^n-Engel and so that $p^n \equiv 1 \pmod{o(b)}$. In the usual way, we get

$$0 = [a + a^{-1}, b^{p^n} + b^{-p^n}] = [a + a^{-1}, b + b^{-1}].$$

Lemma 3.2.2 provides four possible values for ab.

First, we may have $ab = ba$. If $ab = ba^{-1}$, then Lemma 3.2.3 says that either $ab = ba$ or $\langle a, b \rangle \simeq Q_8$. Similarly if $ab = b^{-1}a$. Finally, if $ab = b^{-1}a^{-1}$, then $(ab)^2 = 1$, hence ab is central, by Lemma 3.2.1. That is, $bab = ab^2$, hence $ba = ab$.

Thus, a and b commute or generate a group isomorphic to Q_8. Therefore, the 2-elements form a subgroup that is either abelian or Hamiltonian. As G contains Q_8, it must be Hamiltonian. $\square$

Lemma 3.2.17. *Suppose char $F = p > 2$ and G is a group containing Q_8. If $(FG)^+$ is bounded Lie Engel, then $G \simeq Q_8 \times E \times P$, where E is an elementary abelian 2-group and P is a p-group.*

Proof. Let $Q_8 = \langle g, h \rangle$. Take $c \in G$ such that either c has infinite order or it has odd p'-order. By Lemma 3.2.15, c centralizes $\langle g, h \rangle$. Thus, we have a direct product $\langle g, h \rangle \times \langle c \rangle$. But then by Lemma 3.2.14, $c = 1$. Thus, G is torsion and every element has order $2^i p^j$ for some i and j.

In view of the last lemma, it will suffice to show that the p-elements of G form a subgroup. Take p-elements $a, b \in G$. By Lemma 3.2.15, $\langle a, b \rangle$ centralizes $\langle g, h \rangle$. But no element of order 4 in $Q_8 \times E$ is central; thus, $\langle a, b \rangle$ contains no elements of order 4, hence Q_8 does not lie in $\langle a, b \rangle$. By Theorem 3.2.8, $\langle a, b \rangle$ is nilpotent, hence, a p-group. We are done. $\square$

The second part of the main result of [65] is

Theorem 3.2.18. *Let F be a field of characteristic $p \neq 2$ and G a group containing Q_8. Then $(FG)^+$ is bounded Lie Engel if and only if*

1. *$p = 0$ and $G \simeq Q_8 \times E$, where E is an elementary abelian 2-group; or*
2. *$p > 2$ and $G \simeq Q_8 \times E \times P$, where E is an elementary abelian 2-group and P is a p-group of bounded exponent having a p-abelian normal subgroup of finite index.*

Proof. Let char $F = p > 2$, and suppose that $(FG)^+$ is bounded Lie Engel. Lemma 3.2.17 tells us that $G \simeq Q_8 \times E \times P$, where E is an elementary abelian 2-group and P is a p-group. By Lemma 3.2.14, P has bounded exponent. Since Q_8 is not a subgroup of P, Theorem 3.2.8 completes the necessity part of the proof.

Let us consider the sufficiency. We have $G = Q_8 \times E \times P$, and let A be the p-abelian normal subgroup of finite index in P. Since $E \times A$ is a p-abelian subgroup of finite index in G, Proposition 1.1.4 tells us that FG satisfies a polynomial identity. Thus, by Lemma 1.3.14, $\Delta(G,P)$ is nil of bounded exponent. Let us say that this exponent is at most p^k.

Now, an element of $(FG)^+$ is a linear combination of terms of the form $gc + g^{-1}c^{-1}$, with $g \in Q_8 \times E$ and $c \in P$. But

$$gc + g^{-1}c^{-1} = g + g^{-1} + g(c-1) + g^{-1}(c^{-1}-1).$$

By Lemma 2.1.1, $g + g^{-1}$ is central in $F(Q_8 \times E)$, hence in FG. Also, $g(c-1) + g^{-1}(c^{-1}-1) \in \Delta(G,P)$. Thus, every element of $(FG)^+$ can be written in the form $\gamma + \delta$, with γ central and $\delta \in \Delta(G,P)$. Then $(\gamma + \delta)^{p^k} = \gamma^{p^k} + \delta^{p^k} = \gamma^{p^k}$, and this is central. In particular, for any $\alpha, \beta \in (FG)^+$, we have

$$[\alpha, \underbrace{\beta, \ldots, \beta}_{p^k \text{ times}}] = [\alpha, \beta^{p^k}] = 0,$$

since β^{p^k} is central. That is, $(FG)^+$ is Lie p^k-Engel, as required.

Finally, suppose that char $F = 0$. If $(FG)^+$ is bounded Lie Engel, then by reducing to the prime characteristic case, $G \simeq Q_8 \times E \times P$, where P is simultaneously a 3-group and a 5-group, hence $P = 1$. Conversely, if G is a Hamiltonian 2-group, then by Lemma 2.1.1, the symmetric elements of FG commute. We are done. $\square$

3.3 Lie Nilpotence

We now move on to group rings whose symmetric elements are Lie nilpotent. Much of the work, of course, has already been done in the previous section. The first main result, from Giambruno and Sehgal [36] (for groups without 2-elements) and Lee [64] (for groups with 2-elements) is the following.

Theorem 3.3.1. *Let F be a field of characteristic $p \neq 2$ and G a group not containing the quaternions. Then the following are equivalent:*

 (i) $(FG)^+$ is Lie nilpotent;
 (ii) FG is Lie nilpotent;
(iii) G is nilpotent and p-abelian.

Of course, it is obvious that (ii) implies (i). Let us show that (i) implies (iii). Suppose that FG is Lie nilpotent. Lemma 3.2.5 completely resolves the characteristic zero case. If char $F = p > 2$, then we see from Theorem 3.2.8 that G is nilpotent and, in view of Lemma 3.2.5 and Proposition 1.3.7, that G' is a p-group of bounded exponent. Thus, we only need to verify that G' is finite.

Our proof is by induction on the nilpotency class of G. If G is abelian there is nothing to do. Otherwise, by our inductive hypothesis, $(G/\zeta(G))'$ is a finite p-group. But

$$(G/\zeta(G))' = G'\zeta(G)/\zeta(G) \simeq G'/(G' \cap \zeta(G)).$$

If $G' \cap \zeta(G)$ is finite, then G' is finite, as required. If, on the other hand, $G' \cap \zeta(G)$ is infinite, then G contains an infinite central p-subgroup. But as the $*$-polynomial identity

$$[x_1 + x_1^*, \ldots, x_n + x_n^*] = 0$$

is multilinear, we apply Lemma 3.2.9 to each variable in turn, and obtain

Lemma 3.3.2. *Let F be any field and G a group such that the set $\{g^2 : g \in \zeta(G)\}$ is infinite. If $(FG)^+$ satisfies the identity $[x_1, \ldots, x_n] = 0$, then FG satisfies the identity $[x_1, \ldots, x_n] = 0$.*

As Theorem 3.1.1 tells us that (ii) and (iii) are equivalent, we are done. However, it will pay dividends later if we prove a slightly stronger statement now.

Let R be any ring and S a subset of R. We let $S^{(1)} = R$, and then for each $i \geq 2$, let $S^{(i)}$ be the (associative) ideal of R generated by all elements of the form $[a, b]$, with $a \in S^{(i-1)}$, $b \in S$. We say that S is strongly Lie nilpotent if $S^{(i)} = 0$ for some i. Clearly, strong Lie nilpotence implies Lie nilpotence. If $R = S = FG$, however, the concepts are equivalent, as we can see from the following two lemmas that complete the proof of Theorem 3.3.1.

Lemma 3.3.3. *Let R be a ring and S a subset of R. Suppose, for some $i \geq 1$, that $S^{(i)} \subseteq zR$, where z is central in R. Then for all $j > 0$, we have $S^{(i+j)} \subseteq zS^{(j)}$. In particular, for any positive integer m, $S^{(mi)} \subseteq z^m R$.*

Proof. Our proof is by induction on j. If $j = 1$, then since $S^{(i+1)} \subseteq S^{(i)}$, there is nothing to do. Suppose that $S^{(i+j)} \subseteq zS^{(j)}$. Take any $a \in S^{(i+j)}$, $b \in S$. Then $a = za_1$ for some $a_1 \in S^{(j)}$. Thus, $[a, b] = z[a_1, b] \in zS^{(j+1)}$, as required.

To get the second part, we notice that $S^{(2i)} \subseteq zS^{(i)} \subseteq z^2 R$, and so forth.   □

Lemma 3.3.4. *Let F be a field of characteristic $p > 0$ and let G be a nilpotent group such that $|G'| = p^k$. Then $(FG)^{(2p^k)} = 0$.*

Proof. Our proof is by induction on k. If $k = 0$, then G is abelian, hence $(FG)^{(2)} = 0$, as required. Otherwise, since G is nilpotent, we may take an element z of order p in $G' \cap \zeta(G)$. Then $(G/\langle z \rangle)' = G'/\langle z \rangle$. Letting $\bar{G} = G/\langle z \rangle$, our inductive hypothesis tells us that $(F\bar{G})^{(2p^{k-1})} = 0$. That is,

$$(FG)^{(2p^{k-1})} \subseteq \Delta(G, \langle z \rangle) = (z - 1)FG.$$

As z is central, we apply the last lemma to get $(FG)^{(2p^k)} \subseteq (z - 1)^p FG = 0$, as required.   □

It should be noted that the number $2p^k$ in the last lemma is by no means minimal. We could certainly replace $S^{(i+j)}$ with $S^{(i+j-1)}$ in Lemma 3.3.3 and proceed accordingly, but we are only concerned about finding a bound.

The author also classified the groups G containing Q_8 such that $(FG)^+$ is Lie nilpotent, in [64]. Once again, a stronger version of the sufficiency will be helpful later.

Lemma 3.3.5. *Let char $F = p > 2$ and let $G = Q_8 \times E \times P$, where E is an elementary abelian 2-group and P is a finite p-group. Then $(FG)^+$ is strongly Lie nilpotent.*

Proof. We claim that $((FG)^+)^{(2|P|)} = 0$. Our proof is by induction on $|P|$. If $P = 1$, then by Lemma 2.1.1, the symmetric elements in FG are central, hence $((FG)^+)^{(2)} = 0$, as required. Suppose $|P| = p^m > 1$. Choosing $z \in \zeta(P)$ with $o(z) = p$, we now apply our inductive hypothesis to $\bar{G} = G/\langle z \rangle$. Then $((F\bar{G})^+)^{(2p^{m-1})} = 0$. Thus, $((FG)^+)^{(2p^{m-1})} \subseteq \Delta(G, \langle z \rangle) = (z-1)FG$. By Lemma 3.3.3, $((FG)^+)^{(2p^m)} \subseteq (z-1)^p FG = 0$, as required. $\square$

The second part of the main result of [64] is

Theorem 3.3.6. *Let F be a field of characteristic $p \neq 2$ and G a group containing Q_8. Then $(FG)^+$ is Lie nilpotent if and only if*

1. *$p = 0$ and $G \simeq Q_8 \times E$, where E is an elementary abelian 2-group; or*
2. *$p > 2$ and $G \simeq Q_8 \times E \times P$, where E is an elementary abelian 2-group and P is a finite p-group.*

Proof. If $p = 0$, then Theorem 3.2.18 and Lemma 2.1.1 do the job. So, assume that $p > 2$ and $(FG)^+$ is Lie nilpotent. In view of Theorem 3.2.18, we have $G = Q_8 \times E \times P$, where P is a p-group. We need only show that P is finite. Suppose P is infinite. Since P does not contain Q_8, Theorem 3.3.1 says that P' is finite. Now, $(F(G/P'))^+$ is Lie nilpotent, but $G/P' = Q_8 \times E \times (P/P')$ has an infinite central p-subgroup. Thus, by Lemma 3.3.2, $F(G/P')$ is Lie nilpotent. But G/P' has Q_8 as a subgroup, contradicting Theorem 3.1.1. The proof of the necessity is complete. As the sufficiency follows from Lemma 3.3.5, we are done. $\square$

Remark 3.3.7. Giambruno and Sehgal also showed in [36] that if G has no 2-elements and $(FG)^-$ is Lie nilpotent, then FG is Lie nilpotent. Work on groups containing 2-elements was begun in Giambruno and Polcino Milies [32] and completed in Giambruno and Sehgal [37].

3.4 Lie Solvability

We now present results on the Lie solvability of the symmetric elements in a group ring. These are from Lee et al. [71] and Lee and Spinelli [73]. We assume that char $F \neq 2$ and G has no 2-elements. (No result is currently known for groups with 2-elements.) Our first theorem deals with the characteristic zero case and two different prime characteristic cases.

Theorem 3.4.1. *Let F be a field of characteristic $p \neq 2$ and G a group containing no 2-elements. Suppose either that $p = 0$ or else $p > 2$ and either*

1. *G has only finitely many p-elements, or*
2. *G contains an element of infinite order.*

Then the following are equivalent:

(i) *$(FG)^+$ is Lie solvable;*
(ii) *FG is Lie solvable;*
(iii) *G is p-abelian.*

As Theorem 3.1.3 says that (ii) and (iii) are equivalent, and it is obvious that (ii) implies (i), it remains only to verify that (i) implies (iii).

Notice that for fields of characteristic different from 2, $(FG)^+$ is Lie solvable if and only if there exists an n such that $[g_1 + g_1^{-1}, \ldots, g_{2^n} + g_{2^n}^{-1}]^o = 0$ for all $g_i \in G$. Thus, only the characteristic of the field is relevant, and we are free to assume that F is algebraically closed where convenient. We need to borrow the following result.

Proposition 3.4.2. *Let F be an infinite field of characteristic different from 2 and, for some positive integer n, let $*$ be an involution on $M_n(F)$ fixing F elementwise. Then $M_n(F)$ satisfies the same $*$-polynomial identities as $M_n(F)$ under either the transpose or symplectic involution (with the latter occurring only if n is even).*

Proof. See [41, Theorem 3.6.8]. □

If R is a ring and S is any subset of R, then we let $\delta^{[0]}(S) = S$ and we let $\delta^{[i+1]}(S)$ be the additive subgroup of R generated by $[s_1, s_2]$, as s_1, s_2 run through $\delta^{[i]}(S)$. Thus, S is Lie solvable if and only if $\delta^{[i]}(S) = 0$ for some i. We begin with

Lemma 3.4.3. *Let F be an infinite field of characteristic different from 2 and let $n \geq 3$. If $*$ is any involution on $M_n(F)$ fixing F elementwise, then the matrices symmetric with respect to $*$ are not Lie solvable.*

Proof. Assume that $(M_n(F))^+$ is Lie solvable. In view of Proposition 3.4.2, we may assume that $*$ is either the transpose involution or the symplectic involution.

First, let $*$ be the transpose involution. Then for any $i \neq j$, we have $E_{ij} + E_{ji} \in (M_n(F))^+$. Let i, j, k be distinct. Then notice that

$$[E_{ki} + E_{ik}, E_{kj} + E_{jk}] = E_{ij} - E_{ji},$$

hence

$$E_{ij} - E_{ji} \in \delta^{[1]}((M_n(F))^+)$$

for all $i \neq j$. However, for any distinct i, j, k, we have

$$[E_{kj} - E_{jk}, E_{ki} - E_{ik}] = E_{ij} - E_{ji}.$$

Thus, we see that

$$E_{ij} - E_{ji} \in \delta^{[2]}((M_n(F))^+)$$

for all $i \neq j$, and therefore, repeating this procedure, we get

$$E_{ij} - E_{ji} \in \delta^{[r]}((M_n(F))^+)$$

for all $r \geq 1$. Thus, $(M_n(F))^+$ is not Lie solvable.

On the other hand, suppose that $*$ is the symplectic involution. Then $n = 2m$. Choose distinct i and j with $1 \leq i, j \leq m$. Then

$$E_{ij} + E_{j+m,i+m}, E_{i,j+m} - E_{j,i+m}, E_{i+m,j} - E_{j+m,i} \in (M_n(F))^+.$$

Hence

$$2E_{i,i+m} = [E_{i,j+m} - E_{j,i+m}, E_{ij} + E_{j+m,i+m}] \in \delta^{[1]}((M_n(F))^+),$$

$$2E_{j+m,j} = [E_{ij} + E_{j+m,i+m}, E_{i+m,j} - E_{j+m,i}] \in \delta^{[1]}((M_n(F))^+)$$

and

$$E_{ii} + E_{j+m,j+m} - E_{jj} - E_{i+m,i+m} = [E_{ij} + E_{j+m,i+m}, E_{ji} + E_{i+m,j+m}]$$
$$\in \delta^{[1]}((M_n(F))^+).$$

However, we can see that

$$2E_{i,i+m} = [E_{ii} + E_{j+m,j+m} - E_{jj} - E_{i+m,i+m}, E_{i,i+m}],$$

$$2E_{j+m,j} = [E_{ii} + E_{j+m,j+m} - E_{jj} - E_{i+m,i+m}, E_{j+m,j}]$$

and

$$E_{ii} - E_{i+m,i+m} - (E_{jj} - E_{j+m,j+m}) = [E_{i,i+m}, E_{i+m,i}] - [E_{j,j+m}, E_{j+m,j}].$$

Thus, for every $r \geq 1$, $\delta^{[r]}((M_n(F))^+)$ contains $E_{i,i+m}$, $E_{i+m,i}$ and $E_{ii} - E_{i+m,i+m} - E_{jj} + E_{j+m,j+m}$ for all distinct i, j with $1 \leq i, j \leq m$. In particular, $(M_n(F))^+$ is not Lie solvable.   □

Lemma 3.4.4. *Let F be an algebraically closed field and A a finite-dimensional semisimple F-algebra. Suppose that A has an involution fixing F elementwise. If A^+ is Lie solvable, then the Wedderburn decomposition of A is*

$$F \oplus \cdots \oplus F \oplus M_2(F) \oplus \cdots \oplus M_2(F).$$

Proof. Let $e_1, \ldots, e_k$ be the primitive central idempotents of A. We know that $Ae_1 \cong M_n(F)$ for some n. Suppose $e_1^* = e_1$. Then $*$ induces an involution of $M_n(F)$ fixing F elementwise. By the preceding lemma, $n \leq 2$.

On the other hand, suppose $e_1^* \neq e_1$. Then e_1^* is another primitive central idempotent, say e_2. If $a \in A$, then $ae_1 + a^*e_2 \in A^+$. Thus, every element of $M_n(F)$ is in the image of A^+ under the natural projection. That is, $M_n(F)$ is Lie solvable. But even the symmetric matrices under the transpose involution are not Lie solvable if $n \geq 3$; hence we must have $n \leq 2$, as required.   □

Since the $*$-polynomial identity $[x_1 + x_1^*, \ldots, x_{2^n} + x_{2^n}^*]^o = 0$ is multilinear, we apply Lemma 3.2.9 to each of the indeterminates, and obtain

Lemma 3.4.5. *Let F be any field and G a group such that $\{z^2 : z \in \zeta(G)\}$ is infinite. If $(FG)^+$ satisfies $[x_1, \ldots, x_{2^n}]^o = 0$, then FG satisfies $[x_1, \ldots, x_{2^n}]^o = 0$.*

We also need to assume familiarity with some basic facts about the degrees of irreducible representations.

Proposition 3.4.6. *Let F be an algebraically closed field and G a finite group.*

1. *If every irreducible representation of G over F has degree at most n, then every subgroup and homomorphic image of G inherits this property.*
2. *If char F does not divide $|G|$, then the degree of every irreducible representation of G over F divides $|G|$.*

Proof. The first part is [82, Lemma 6.1.3]. For the second part, the characteristic zero case is [89, 8.3.11]. From the discussion of Brauer characters in [50, p. 268], we see that it follows for fields of prime characteristic as well. $\qquad\square$

We can now prove part of Theorem 3.4.1.

Proposition 3.4.7. *Let char $F = p \neq 2$. Suppose that G has no 2-elements and, if $p > 2$, that G has no p-elements. If $(FG)^+$ is Lie solvable, then G is abelian.*

Proof. By Proposition 2.1.2, FG satisfies a polynomial identity. Thus, by Proposition 1.1.4, G has a p-abelian (hence abelian) normal subgroup A of finite index. Take $g \in G$ and let $H = \langle A, g \rangle$. We claim that H is abelian.

If g has infinite order, then some power of g lies in A and is therefore central in H. By Lemma 3.4.5, FH is Lie solvable. Then Theorem 3.1.3 tells us that H is abelian.

Thus, we may assume that g has finite order. Take $a \in A$. Clearly $a \in \phi(H)$, since A has finite index. Thus, if $o(a) < \infty$, then the normal closure of $\langle a \rangle$ in H is finite. Therefore, $\langle a, g \rangle$ is finite and $F\langle a, g \rangle$ is semisimple. But combining Lemma 3.4.4 and Proposition 3.4.6 and assuming that F is algebraically closed, we see that $F\langle a, g \rangle$ is a direct sum of copies of F. Thus, a and g commute.

Suppose, on the other hand, that a has infinite order. Then the normal closure of $\langle a \rangle$ in H is a finitely generated abelian group. Let it be $L = M \times N$, where M is a direct product of infinite cyclic groups and N is finite. Letting $n = |N|$, we have $L^n = M^n$. Fix any odd prime $q \neq p$, and let $L_k = \langle L/L^{nq^k}, L^{nq^k} g \rangle$, for any positive integer k. We can see that L_k has finite odd order not divisible by p. Also, $(FL_k)^+$ is Lie solvable. As we saw above, L_k must be abelian. Thus, $(g, a) \in L^{nq^k} = M^{nq^k}$ for all $k \geq 1$. But M is a direct product of infinite cyclic groups, hence the intersection of these over all k is trivial. That is, a and g commute.

Thus, we can conclude that A is central in G. If A is infinite, then as A has no 2-elements, Lemma 3.4.5 says that FG is Lie solvable; hence, by Theorem 3.1.3, G is abelian. If A is finite, then so is G. But we dealt with the finite case above. We are done. $\qquad\square$

The characteristic zero case of Theorem 3.4.1 is now complete. Thus, we assume that $p > 2$. The following observation is handy.

Lemma 3.4.8. *Let F be a field of characteristic $p > 2$. Let G be a group having a finite normal p-subgroup N. Then $(FG)^+$ is Lie solvable if and only if $(F(G/N))^+$ is Lie solvable.*

Proof. The necessity is clear. Suppose that $(F(G/N))^+$ is Lie solvable. Choose n so that $\delta^{[n]}((F(G/N))^+) = 0$. Then $\delta^{[n]}((FG)^+) \subseteq \Delta(G,N)$. By Lemma 1.1.1, $\Delta(G,N)$ is a nilpotent ideal. Let us say that $(\Delta(G,N))^{2^k} = 0$. A simple induction reveals that for all $i \geq 0$, $\delta^{[n+i]}((FG)^+) \subseteq (\Delta(G,N))^{2^i}$. Thus, $\delta^{[n+k]}((FG)^+) = 0$, as required.   □

In particular, the case of Theorem 3.4.1 in which G has only finitely many p-elements will be complete once we have shown that the p-elements form a group. We begin with the case where G is finite. Recall that, for any prime q, H is said to be a Hall q'-subgroup of G if H is a q'-group of q-power index in G. A normal Hall q'-subgroup is also known as a normal q-complement. We recall the following result due to Frobenius, which is [89, 10.3.2].

Proposition 3.4.9. *Let G be a finite group and q a prime. Then G has a normal q-complement if and only if every q-subgroup of G is centralized by the q'-elements in its normalizer.*

The next two lemmas are taken from Lee [68]. Write $N_G(H)$ for the normalizer of H in G.

Lemma 3.4.10. *Let F be an algebraically closed field of characteristic $p > 2$ and let G be a finite group such that every irreducible F-representation of G has degree 1 or 2. Assume that G is not a p-group and let q be the smallest prime, other than p, dividing $|G|$. Then G has a normal q-complement.*

Proof. Our proof is by induction on $|G|$. Suppose, first of all, that G has no nontrivial normal q-subgroup. Let $M \neq 1$ be a q-subgroup. Then $N_G(M)$ is a proper subgroup of G. By Proposition 3.4.6, the irreducible representations of $N_G(M)$ have degree at most 2. Thus, by our inductive hypothesis, $N_G(M)$ has a normal q-complement, K. That is, K consists of all of the q'-elements in $N_G(M)$. As M and K are normal subgroups of relatively prime order in $N_G(M)$, they centralize each other. Thus, by Frobenius' theorem, G has a normal q-complement.

Therefore, let us assume that G has a nontrivial normal q-subgroup, N. If N is not abelian, then we may replace it with N'. As N is nilpotent, repeating this will eventually yield a nontrivial abelian normal q-subgroup, N. Since the irreducible representations of G/N have degree at most 2, G/N has a normal q-complement, H/N. Thus, by the Schur–Zassenhaus theorem, $H = N \rtimes L$, where L is a q'-group.

We claim that $H = N \times L$. Let V be a simple right FH-module that is 2-dimensional over F, and let W be a simple FN-submodule of V. As N is abelian, W must be 1-dimensional over F. By Clifford's theorem (see [89, 8.1.3]), for each

$h \in H$, Wh is a simple FN-module and $V = \sum_{h \in H} Wh$. But comparing dimensions shows that $V = W \oplus Wh$ for some $h \in H$.

Suppose that W and Wh are not isomorphic as FN-modules. Then Clifford's theorem says that, for every $g \in L$, $Wg = W$ or Wh, and $Whg = W$ or Wh. That is, L has an action on the set $\{W, Wh\}$. The orbits must have order 1 or 2, and must have order dividing $|L|$. But if $|G|$ is even, then $q = 2$, hence L has odd order in any case. Therefore, the orbits have order 1. Obviously $W1 = W$, hence $WL = W$. But, by definition, $WN = W$. Therefore, $WG = W$, contradicting the irreducibility of V.

Thus, W and Wh are isomorphic FN-modules. As they are one-dimensional, every element of N acts on each of them (and hence on their direct sum, V) as an element of F. This would clearly be true for all one-dimensional modules as well. Thus, the actions of elements of N commute with all other actions on V. In particular, (g, h) acts trivially on V for all $g \in N$, $h \in H$ and all simple FH-modules V (since they all have F-dimension at most 2, by Proposition 3.4.6).

Writing $FH/J(FH) \cong M_{n_1}(F) \oplus \cdots \oplus M_{n_r}(F)$ as in Proposition 1.3.3, we see that $(g, h) - 1 \in J(FH)$, and $J(FH)$ is nilpotent. Thus, (g, h) is a p-element. But $g \in N$, and N is a normal p'-group. That is, $(g, h) = 1$, and N is central in H. In particular, $H = N \times L$, as claimed.

Since L consists of the q'-elements of H, it is normal in G. Furthermore,

$$(G : L) = (G : H)(H : L) = (G/N : H/N)(H : L);$$

hence G/L is a q-group. Thus, L is a normal q-complement in G. $\square$

Lemma 3.4.11. *Let F be an algebraically closed field of characteristic $p > 2$ and G a finite group. If every irreducible representation of G over F has degree 1 or 2, then the p-elements of G form a (normal) subgroup.*

Proof. Our proof is by induction on $|G|$. If G is a p-group, there is nothing to prove. Otherwise, let q be the smallest prime, besides p, dividing $|G|$. By the last lemma, G has a normal q-complement, H. But by Proposition 3.4.6, every irreducible representation of H has degree at most 2. Since $|H| < |G|$, our inductive hypothesis tells us that the p-elements of H form a subgroup. As G/H is a p'-group, all of the p-elements of G lie in H. We are done. $\square$

In terms of Lie solvability, this allows us to prove the following.

Lemma 3.4.12. *Let F be a field of characteristic $p > 2$ and G a group. If $(FG)^+$ is Lie solvable, then the p-elements of G form a (normal) subgroup.*

Proof. We may assume that F is algebraically closed. Suppose, first of all, that G is finite. As we observed in the proof of Lemma 2.3.2, $J(FG)$ is invariant under $*$. Thus, $FG/J(FG)$ has an induced involution. Furthermore, if $\alpha \in FG$ is such that $\alpha + J(FG)$ is symmetric, then $\frac{\alpha + \alpha^*}{2} + J(FG) = \alpha + J(FG)$. Thus, $(FG/J(FG))^+$ is the homomorphic image of $(FG)^+$. In particular, $(FG/J(FG))^+$ is Lie solvable. However, $FG/J(FG)$ is semisimple, and by Lemma 3.4.4,

$$FG/J(FG) \cong F \oplus \cdots \oplus F \oplus M_2(F) \oplus \cdots \oplus M_2(F).$$

That is, the irreducible representations of G have degree at most 2. In view of the previous lemma, we are done.

Next, suppose that G is torsion. By Proposition 2.1.2, FG satisfies a polynomial identity, hence, by Proposition 1.1.4, G is locally finite. If $g, h \in G$ are p-elements, it suffices to show that $\langle g, h \rangle$ is a p-group. That is, it suffices to consider the finite case, which we have already done.

Finally, let G be arbitrary. As above, we may as well assume that $G = \langle g, h \rangle$, where g and h are p-elements. As we just observed, FG satisfies a polynomial identity, so let A be a p-abelian normal subgroup of finite index. If we can show that the p-elements of G/A' form a group, then it is true for G as well. Thus, we can take A to be abelian.

Now, G/A is a finite group; hence, as we saw above, ghA is a p-element, say $(gh)^{p^m} \in A$. Choosing any positive integer n relatively prime to p, we have $(gh)^{p^m n} \in A^n$. But G/A^n is torsion. Hence, by the torsion case, ghA^n is also a p-element. It follows that $(gh)^{p^m} \in A^n$ for every n relatively prime to A. Now, every subgroup of finite index in a finitely generated group is finitely generated (see, for instance, [94, Proposition I.4.1]). Thus, A is a finitely generated abelian group. Let us write $A = A_1 \times A_2 \times A_3$, where A_1 is a finite p-group, A_2 is a finite p'-group and A_3 is a direct product of infinite cyclic groups. The intersection of the A^n, as n runs over the positive integers relatively prime to p, is A_1. That is, gh is a p-element, which completes our proof. □

We can now fill in another piece of Theorem 3.4.1.

Proposition 3.4.13. *Let char $F = p > 2$ and let G be a group containing no 2-elements and finitely many p-elements. Then $(FG)^+$ is Lie solvable if and only if FG is Lie solvable.*

Proof. Suppose that $(FG)^+$ is Lie solvable. By the preceding lemma, the p-elements of G form a subgroup, P. Thus, $(F(G/P))^+$ is Lie solvable. Applying Proposition 3.4.7, we see that G/P is abelian, hence G' is a finite p-group. Theorem 3.1.3 completes the proof. □

Let us consider what remains. Let G be a group without 2-elements. We have handled the characteristic zero case, so let char $F = p > 2$ and let P be the infinite set of p-elements of G, which we know to be a subgroup whenever $(FG)^+$ is Lie solvable. Thus, as $(F(G/P))^+$ is Lie solvable, Proposition 3.4.7 tells us that G/P is abelian, hence G' is a p-group. It remains only to show that G' is finite.

As FG satisfies a polynomial identity, we have a p-abelian normal subgroup A of finite index. By Lemma 3.4.8, $(FG)^+$ is Lie solvable if and only if $(F(G/A'))^+$ is Lie solvable. Thus, we can factor out A' and assume that A is abelian. Let $\{g_1, \ldots, g_r\}$ be a transversal of A in G. For each i, let $H_i = \langle A, g_i \rangle$. We claim that it is sufficient to show that for each i, H_i' is finite. Indeed, every element of G lies in some H_i, and if an element of H_i has at most n conjugates by elements of H_i, then it surely has at most nr conjugates in G. But we recall the famous result of Neumann (see [89, 14.5.11]); which states that the number of conjugates of the elements of a group G is bounded if and only if G' is finite. Thus, if each $\langle A, g_i \rangle'$ is finite, then we obtain a bound on the number of conjugates for all elements of G. Again applying Neumann's theorem, G' is finite, as required.

Thus, we take $G = \langle A, g \rangle$, where A is abelian. If $o(g) = \infty$, then some power of g lies in A and is therefore central. In view of Lemma 3.4.5, FG is Lie solvable, and we are done. Thus, we assume that g has finite order.

The following proposition completes the proof of Theorem 3.4.1.

Proposition 3.4.14. *Let F be a field of characteristic $p > 2$ and G a nontorsion group containing no 2-elements. If $(FG)^+$ is Lie solvable, then FG is Lie solvable.*

Proof. As we saw above, we may assume that $G = \langle A, g \rangle$, where A is an abelian normal subgroup and g has finite order m. Naturally, if G has an element of infinite order, then so does A, so take $a \in A$ of infinite order. We have seen that G' is a p-group, hence $(a, g) \in A \cap P$, where P is the group of p-elements. That is, $a^g \equiv a$ (mod $A \cap P$), and therefore, $a^{g^i} \equiv a$ (mod $A \cap P$) for all positive integers i. It is easy to see that $aa^g a^{g^2} \cdots a^{g^{m-1}}$ lies in A and commutes with g; hence, it is central. But

$$aa^g a^{g^2} \cdots a^{g^{m-1}} \equiv a^m \quad (\text{mod } A \cap P).$$

That is, we have a central element of the form $a^m b$, where a^m has infinite order and $b \in P$. As a and b commute, $a^m b$ is a central element of infinite order. Lemma 3.4.5 completes the proof. $\square$

Thus, we can assume that our group is torsion. No result is known that completely covers the remaining case, but the following theorem, also from Lee et al. [71] and Lee and Spinelli [73], will be useful in discussing the solvability of the symmetric units.

Theorem 3.4.15. *Let F be a field of characteristic $p > 2$. Let G be a group containing an infinite p-subgroup of bounded exponent, but no nontrivial elements of order dividing $p^2 - 1$. Then the following are equivalent:*

(i) $(FG)^+$ is Lie solvable;
(ii) FG is Lie solvable;
(iii) G is p-abelian.

A group-theoretic lemma is required before we present the proof of the theorem.

Lemma 3.4.16. *Let p be any prime and let G be a group having an abelian normal subgroup A of finite index as well as an infinite p-subgroup H of bounded exponent. Then A contains an infinite direct product of nontrivial finite abelian p-groups, each of which is normal in G.*

Proof. We can see that $A \cap H$ is a subgroup of finite index in H; hence, it is infinite. But by the Prüfer–Baer theorem (see [89, 4.3.5]), an infinite abelian p-group of bounded exponent is a direct product of cyclic groups. In particular, then, A has infinitely many elements of order p. Thus, A has an infinite subgroup $B = \prod_{i \in I} B_i$, where B is normal in G and each B_i is cyclic of order p.

Let I_0 be any finite subset of I. We claim that there is a nontrivial finite normal subgroup of G contained in $K = \prod_{i \in I \setminus I_0} B_i$. Indeed, let $\{g_1, \ldots, g_m\}$ be a transversal of A in G containing 1. Then since $(B : K) < \infty$, we get $(B : g_j K g_j^{-1}) < \infty$. Thus, letting $N = \bigcap_{j=1}^{m} g_j K g_j^{-1}$, we have $(B : N) < \infty$. Take any $1 \neq g \in N$. Then g commutes with all $a \in A$, and $g_j^{-1} g g_j \in K$ for all j. Thus, every conjugate of g lies in K, which is an abelian p-group. Therefore, the normal closure of $\langle g \rangle$ in G is a finite subgroup of K. The claim is proved.

Taking I_0 to be the empty set, we construct a nontrivial finite normal subgroup A_1 of G contained in $\prod_{i \in I} B_i$. Thus, there is a finite set I_1 such that $A_1 \leq \prod_{i \in I_1} B_i$. Applying the same procedure with I_1 in place of I_0, we get a nontrivial finite normal subgroup A_2 of G contained in $\prod_{i \in I \setminus I_1} B_i$. We can then find a finite subset I_2 of I such that $A_1 \times A_2 \leq \prod_{i \in I_2} B_i$, and so forth. In this way, we get the desired direct product of finite normal subgroups A_i. $\qquad\square$

Proof of Theorem 3.4.15. We only need to verify that (i) implies (iii). It will suffice to take $F = \mathbb{Z}_p$. We have seen that we can assume that G is torsion and has an abelian normal subgroup A as well as an element g such that $G = \langle A, g \rangle$. In addition, A contains an infinite direct product $A_1 \times A_2 \times \cdots$ of nontrivial finite abelian p-subgroups, each of which is normal in G. Furthermore, we know that $G' \leq P$, where P is the subgroup of p-elements. It remains only to show that G' is finite.

Let N be the set of p'-elements of A. Then N is a normal subgroup of G. Furthermore, $(G, N) \leq G' \cap N \leq P \cap N = 1$. Thus, N is central. If N is infinite, then so is N^2, as it contains no 2-elements. Thus, in view of Lemma 3.4.5, FG is Lie solvable, and we are done. Therefore, we may assume that N is finite. But then $(F(G/N))^+$ is Lie solvable, and $(G/N)'$ is finite if and only if G' is finite. Thus, we factor out N and let A be a p-group.

Write $g = hk$, where h is a p-element and k a p'-element. Then each $\langle A_i, h \rangle$ is a finite p-group. As A_i is a normal subgroup, it contains a nonidentity element central in $\langle A_i, h \rangle$, hence in $\langle A, h \rangle$. Thus, $\langle A, h \rangle$ has an infinite central p-subgroup. By Lemma 3.4.5, $F\langle A, h \rangle$ is Lie solvable, hence $\langle A, h \rangle'$ is finite. Of course, $G' \leq A$, hence $\langle A, h \rangle$ is normal in G as well. Thus, let us factor out $\langle A, h \rangle'$ and assume that $\langle A, h \rangle$ is abelian. Replacing A with $\langle A, h \rangle$, we have $G = \langle A, k \rangle$, Therefore, we may assume that g is a p'-element. Let us write $o(g) = q$. Thus, $G = A \rtimes \langle g \rangle$.

If g centralizes infinitely many of the A_i, then G has an infinite central p-subgroup; hence, by Lemma 3.4.5, FG is Lie solvable. Thus, let us discard the finitely many A_i that are centralized by g and assume that for each $i \geq 1$, there exists $a_i \in A_i$ such that $a_i^g \neq a_i$.

We wish to show that $\delta^{[i]}((FG)^+) \neq 0$ for all $i \geq 0$. To this end, let

$$\alpha_0 = a_1 + a_1^{-1} - a_1^g - a_1^{-g},$$

where $a_1^{-g} = (a_1^{-1})^g$. Then for each $i \geq 0$, let

$$\alpha_{i+1} = \alpha_i^{g^{2^i}} \beta_i - \alpha_i \beta_i^{g^{2^i}}$$

where, for every i, β_i is obtained by replacing

$$a_1, a_2, \ldots, a_{2^i}$$

in α_i with

$$a_{2^i+1}, a_{2^i+2}, \ldots, a_{2^{i+1}},$$

respectively.

We claim that, for each $i \geq 0$, $\delta^{[i+1]}((FG)^+)$ contains $g^{2^i} \alpha_i - g^{-2^i} \alpha_i^{g^{-2^i}}$. To verify the $i = 0$ case, we note that

$$[g + g^{-1}, a_1 + a_1^{-1}] = g(a_1 + a_1^{-1} - a_1^g - a_1^{-g}) + g^{-1}(a_1 + a_1^{-1} - a_1^{g^{-1}} - a_1^{-g^{-1}})$$

$$= g\alpha_0 - g^{-1}\alpha_0^{g^{-1}}.$$

Assume now that $g^{2^i} \alpha_i - g^{-2^i} \alpha_i^{g^{-2^i}} \in \delta^{[i+1]}((FG)^+)$ (and similarly for $g^{2^i} \beta_i - g^{-2^i} \beta_i^{g^{-2^i}}$). Noting that the α_i and β_i, as well as their conjugates, commute (since they are in FA),we obtain

$$[g^{2^i} \alpha_i - g^{-2^i} \alpha_i^{g^{-2^i}}, g^{2^i} \beta_i - g^{-2^i} \beta_i^{g^{-2^i}}] = g^{2^{i+1}} (\alpha_i^{g^{2^i}} \beta_i - \alpha_i \beta_i^{g^{2^i}})$$

$$+ g^{-2^{i+1}} (\alpha_i^{g^{-2^{i+1}}} \beta_i^{g^{-2^i}} - \alpha_i^{g^{-2^i}} \beta_i^{g^{-2^{i+1}}})$$

$$- \alpha_i^{g^{-2^i}} \beta_i^{g^{-2^i}} - \alpha_i \beta_i + \alpha_i \beta_i + \alpha_i^{g^{-2^i}} \beta_i^{g^{-2^i}}$$

$$= g^{2^{i+1}} \alpha_{i+1} - g^{-2^{i+1}} \alpha_{i+1}^{g^{-2^{i+1}}},$$

and the claim is proved.

Since g has odd order, the sets $g^{2^i} A$ and $g^{-2^i} A$ do not intersect. Thus, we will be done if we show that $\alpha_i \neq 0$ for all i. It will simplify matters to make a slightly stronger hypothesis for our induction. We claim that for all $i \geq 0$,

$$\alpha_{i+1} \neq 0, \ \beta_{i+1} \neq 0, \ \alpha_i^g \neq \alpha_i, \ \text{and} \ \beta_i^g \neq \beta_i.$$

Of course, the proof for the β_i is identical to that for the α_i, so we will prove the claim only for the α_i.

Consider the $i = 0$ case. Suppose that $\alpha_0^g = \alpha_0$. Then

$$a_1^g + a_1^{-g} - a_1^{g^2} - a_1^{-g^2} = a_1 + a_1^{-1} - a_1^g - a_1^{-g}.$$

Thus, $a_1^{g^2} \in \{a_1^g, a_1^{-g}, a_1^{-g^2}, a_1, a_1^{-1}\}$. If $a_1^{g^2} = a_1^g$, then $a_1^g = a_1$, contradicting the choice of a_1. Suppose that $a_1^{g^2} = a_1^{-g}$. Then $a_1^g = a_1^{-1}$, hence g^2 centralizes a_1. But g has odd order, and therefore $a_1^g = a_1$, which is impossible. If $a_1^{g^2} = a_1^{-g^2}$, then $a_1^2 = 1$, which gives a contradiction. The cases $a_1^{g^2} = a_1$ and $a_1^{g^2} = a_1^{-1}$ similarly result in contradictions. Thus, $\alpha_0^g \neq \alpha_0$ (and similarly, $\beta_0^g \neq \beta_0$).

We must also verify that $\alpha_1 \neq 0$. But if $\alpha_1 = 0$, then $\alpha_0^g \beta_0 = \alpha_0 \beta_0^g$. Now, $\alpha_0, \alpha_0^g \in FA_1$ and $\beta_0, \beta_0^g \in FA_2$, and the product $A_1 \times A_2$ is direct. The only way this can happen is if either $\beta_0 = 0$ (which would imply that $\beta_0^g = \beta_0$, and we know this is not the case) or $\alpha_0^g = \lambda \alpha_0$ for some $\lambda \in F$. Certainly $\alpha_0 \neq 0$, hence $\lambda \neq 0$. But $\lambda \in F = \mathbb{Z}_p$, hence $\lambda^{p-1} = 1$. Also, $\alpha_0 = \alpha_0^{g^q} = \lambda^q \alpha_0$. Thus, $\lambda^q = 1$. But by our restriction on the orders of the elements of G, $(q, p-1) = 1$. Thus, $\lambda = 1$ and $\alpha_0^g = \alpha$, which we know is not the case. The $i = 0$ step is complete.

Now, assume that for all j, $0 \leq j \leq i$, we have $\alpha_{j+1} \neq 0$, $\beta_{j+1} \neq 0$, $\alpha_j^g \neq \alpha_j$ and $\beta_j^g \neq \beta_j$. Suppose that $\alpha_{i+1}^g = \alpha_{i+1}$. Then $\alpha_{i+1}^{g^{2^i}} = \alpha_{i+1}$, hence

$$(\alpha_i^{g^{2^i}} \beta_i - \alpha_i \beta_i^{g^{2^i}})^{g^{2^i}} = \alpha_i^{g^{2^i}} \beta_i - \alpha_i \beta_i^{g^{2^i}}.$$

Rearranging the terms, we get

$$(\alpha_i^{g^{2^{i+1}}} + \alpha_i)\beta_i^{g^{2^i}} = \alpha_i^{g^{2^i}}(\beta_i + \beta_i^{g^{2^{i+1}}}).$$

On each side of this last equation, we have a product of an element of $F(A_1 \times \cdots \times A_{2^i})$ with an element of $F(A_{2^i+1} \times \cdots \times A_{2^{i+1}})$. As the product $A_1 \times \cdots \times A_{2^{i+1}}$ is direct, and we know by induction that $\beta_i \neq 0$, this implies that

$$\alpha_i^{g^{2^{i+1}}} + \alpha_i = \mu \alpha_i^{g^{2^i}}$$

for some $\mu \in F$.

Let $y = g^{2^i}$, and let V be the F-vector space with basis $\{\alpha_i, \alpha_i^y\}$. (Notice that α_i and α_i^y must be linearly independent. Indeed, if α_i^y is a scalar multiple of α_i, then since g has odd order, $\alpha_i^g = \lambda' \alpha_i$ for some $0 \neq \lambda' \in F$. Thus, $\alpha_i = \alpha_i^{g^q} = (\lambda')^q \alpha_i$ and therefore, $(\lambda')^q = 1$. But $\lambda' \in F$, hence $(\lambda')^{p-1} = 1$. Since $(q, p-1) = 1$, we have $\lambda' = 1$, hence $\alpha_i^g = \alpha_i$, contradicting our inductive hypothesis.) Also notice that $\alpha_i^y \in V$ and $(\alpha_i^y)^y = \alpha_i^{y^2} = \mu \alpha_i^y - \alpha_i \in V$. Thus, y acts as a linear transformation on V.

Write Y for this linear transformation, and let $f(x)$ be its minimal polynomial. Then as $\dim V = 2$, we have $\deg(f) \leq 2$. Let F' be a splitting field for f over F. Then F' is a field extension of degree at most 2 over $\mathbb{Z}_p$; hence, it is a field of p or p^2 elements. Let v be an eigenvalue for Y in F'. Obviously, Y is invertible, so $0 \neq v \in F'$, hence $v^{p^2-1} = 1$. But as $g^q = 1$, we must also have $Y^q = I$. Therefore, $v^q = 1$ as well. Since $(q, p^2 - 1) = 1$, we have $v = 1$.

As 1 is the only eigenvalue for Y, choosing a suitable basis for V, we can regard Y as the matrix

$$\begin{pmatrix} 1 & \mu' \\ 0 & 1 \end{pmatrix}$$

for some $\mu' \in F$. Since $Y^q = I$, we get $q\mu' = 0$. But $(q, p) = 1$, hence $\mu' = 0$, and Y is the identity transformation. It follows that $\alpha_i^{g^{2i}} = \alpha_i$. But g has odd order, hence $\alpha_i^g = \alpha_i$, contradicting our inductive hypothesis. Therefore, $\alpha_{i+1}^g \neq \alpha_{i+1}$ and similarly, $\beta_{i+1}^g \neq \beta_{i+1}$.

Finally, let us suppose that $\alpha_{i+2} = 0$. Then

$$\alpha_{i+1}^{g^{2^{i+1}}} \beta_{i+1} = \alpha_{i+1} \beta_{i+1}^{g^{2^{i+1}}} .$$

On each side of the equation we have a product of an element of $F(A_1 \times \cdots \times A_{2^{i+1}})$ with an element of $F(A_{2^{i+1}+1} \times \cdots \times A_{2^{i+2}})$. By our inductive hypothesis, $\alpha_{i+1} \neq 0 \neq \beta_{i+1}$; hence $\alpha_{i+1}^{g^{2^{i+1}}}$ is a scalar multiple of α_{i+1}. As g has odd order, we deduce that $\alpha_{i+1}^g = \kappa\alpha_{i+1}$ for some $0 \neq \kappa \in F$. But since $g^q = 0$, this implies that $\kappa^q = 1$. However, $\kappa^{p-1} = 1$ as well, hence $\kappa = 1$ and $\alpha_{i+1}^g = \alpha_{i+1}$, giving us a contradiction. Our proof is complete. □

Remark 3.4.17. Since $[(FG)^+, (FG)^+] \subseteq (FG)^-$, it follows that if $(FG)^-$ is Lie solvable, then so is $(FG)^+$. Thus, under any of the conditions upon G imposed in our theorems in this section, if $(FG)^-$ is Lie solvable, then FG is Lie solvable. In fact, if char $F = 0$ or char $F = p > 2$ and G has only finitely many p-elements, Lee et al. [71] classified the groups G containing 2-elements such that $(FG)^-$ is Lie solvable, as well.

They also observed that in order to remove the condition in Theorem 3.4.15 that G contain an infinite p-subgroup of bounded exponent, it is sufficient to consider the case in which G has a normal subgroup A that is a direct product of finitely many copies of the quasicyclic p-group, C_{p^∞}, and $G/A = \langle Ag \rangle$, where the order of g is a prime power. This case, however, remains open. Indeed, the restriction can be dropped whenever G does not have C_{p^∞} as a subhomomorphic image.

Of course, $G = Q_8$ is a group such that $(FG)^+$ is Lie solvable (being commutative), but FG is not. Unfortunately, the usual criterion that G not contain Q_8 will not be sufficient. Indeed, we observed in Remark 3.2.13 that if G is a dihedral group, then $(FG)^-$ is commutative; hence, $(FG)^+$ is Lie solvable. However, it seems reasonable to conjecture that if G has no 2-elements, then $(FG)^+$ is Lie solvable if and only if FG is Lie solvable.

Chapter 4
Nilpotence of $\mathcal{U}(FG)$ and $\mathcal{U}^+(FG)$

4.1 Introduction

We now turn our attention to specific group identities. In particular, we can ask
when, for a field F and a group G, $\mathcal{U}(FG)$ is a nilpotent group. This question was
answered in a series of classical papers. Bateman and Coleman [5] determined the
answer when G is finite. Khripta [56] extended this to groups of characteristic $p > 0$
having at least one p-element. This is a very useful feature, as we can see from the
following lemma.

Lemma 4.1.1. *Let R be a ring of prime characteristic p. Suppose $\eta \in R$ is central
and square-zero. Take $a \in R$ and $u \in \mathcal{U}(R)$. Then we have*

$$(1 + \eta a, \underbrace{u, \ldots, u}_{p^n \; times}) = 1 + \eta(a^{u^{p^n}} - a)$$

for any $n \geq 0$.

Proof. First of all, notice that $1 + \eta a \in \mathcal{U}(R)$, since $(1 + \eta a)^{-1} = 1 - \eta a$. We claim
that, for every positive integer m,

$$(1 + \eta a, \underbrace{u, \ldots, u}_{m \; times}) = 1 + \eta \sum_{i=0}^{m} (-1)^{m-i} \binom{m}{i} a^{u^i}.$$

When $m = 1$, we have

$$(1 + \eta a, u) = (1 - \eta a)u^{-1}(1 + \eta a)u = 1 + \eta(a^u - a),$$

as required. Assuming that the claim holds for m, we get

G.T. Lee, *Group Identities on Units and Symmetric Units of Group Rings*,
Algebra and Applications 12, DOI 10.1007/978-1-84996-504-0_4,
© Springer-Verlag London Limited 2010

$$(1 + \eta a, \underbrace{u, \ldots, u}_{m+1 \text{ times}})$$

$$= \left(1 + \eta \sum_{i=0}^{m} (-1)^{m-i} \binom{m}{i} a^{u^i}, u \right)$$

$$= 1 + \eta \left(a^{u^{m+1}} + (-1)^{m+1} a + \sum_{i=0}^{m-1} (-1)^{m-i} \left(\binom{m}{i} + \binom{m}{i+1} \right) a^{u^{i+1}} \right)$$

$$= 1 + \eta \sum_{i=0}^{m+1} (-1)^{m+1-i} \binom{m+1}{i} a^{u^i},$$

and the claim is proved.

Substituting $m = p^n$, and noting that p divides $\binom{p^n}{i}$ whenever $0 < i < p^n$, we obtain our result.   □

In particular, if $z \in G$ is a central element of order p, then we can use $\eta = \hat{z}$ in the above lemma.

The semiprime case has a different solution. Work on this case was conducted independently by Fisher et al. [30] and Khripta [55].

The determination of the conditions under which $\mathcal{U}^+(FG)$ is nilpotent came much later. As the symmetric units do not, in general, form a group, let us state that we mean specifically that $\mathcal{U}^+(FG)$ satisfies a group identity of the form

$$(x_1, \ldots, x_n) = 1$$

for some $n \geq 2$. In fact, we can just as easily specify that the subgroup of $\mathcal{U}(FG)$ generated by the symmetric units is nilpotent, due to the next result.

Lemma 4.1.2. *Let G be any group and S a subset of G. If S satisfies $(x_1, \ldots, x_n) = 1$, then so does $\langle S \rangle$.*

Proof. We may as well assume that $G = \langle S \rangle$. Our proof is by induction on n. If $n = 2$, then the generators of G commute, hence G is abelian. Otherwise, consider the $n + 1$ case. Let $\bar{G} = G/\zeta(G)$. Then $\bar{G}$ is generated by $\{\bar{g} : g \in S\}$. Also, for any $g_1, \ldots, g_n \in S$, we see that $(g_1, \ldots, g_n)$ commutes with every generator of G, hence $(\bar{g}_1, \ldots, \bar{g}_n) = 1$. By our inductive hypothesis, $\bar{G}$ satisfies $(x_1, \ldots, x_n) = 1$. In particular, $(h_1, \ldots, h_n) \in \zeta(G)$ for all $h_i \in G$, hence $(h_1, \ldots, h_{n+1}) = 1$ for all $h_i \in G$. We are done.   □

Let F be a field of characteristic different from 2 and G a torsion group. In [67], the author found the conditions under which $\mathcal{U}^+(FG)$ is nilpotent. It turns out that the field is only relevant in terms of its characteristic. Thus, the proof does not depend upon the results of Chapter 2, where the field was assumed to be infinite. This turns out to be crucial for our proofs in the next chapter.

On the other hand, if G is not torsion, then we will need to assume that F is infinite. In this case, Lee et al. [69] determined when $\mathcal{U}^+(FG)$ is nilpotent. Here, we must make the usual assumption for the sufficiency that G/T is a u.p. group, where T is the set of torsion elements. (Naturally, this is not necessary in discussing $\mathcal{U}(FG)$, since there we get for free that G is nilpotent, and every torsion-free nilpotent group is a u.p. group.)

In the next section, we will discuss when $\mathcal{U}(FG)$ is nilpotent. The rest of the chapter is devoted to determining the groups G such that $\mathcal{U}^+(FG)$ is nilpotent.

4.2 Nilpotent Unit Groups

Let us first consider the case in which FG is not semiprime. The main result of Khripta [56] is the following.

Theorem 4.2.1. *Let F be a field of characteristic $p > 0$ and G a group containing a p-element. Then $\mathcal{U}(FG)$ is nilpotent if and only if G is nilpotent and p-abelian.*

A few lemmas are required. We begin with

Lemma 4.2.2. *Suppose char $F = p > 0$ and let G be a group containing a central element z of order p. If $\mathcal{U}(FG)$ is p^n-Engel, then $G^{p^{n+1}} \subseteq \zeta(G)$.*

Proof. Take $g, h \in G$. Then as $\eta = \hat{z}$ is central and square-zero, Lemma 4.1.1 tells us that

$$1 = (1 + \eta g, \underbrace{h, \ldots, h}_{p^n \text{ times}}) = 1 + \eta(g^{h^{p^n}} - g).$$

That is, $\eta g^{h^{p^n}} = \eta g$, hence $g^{h^{p^n}} = z^i g$ for some $i \geq 0$. But z is central, hence $g^{h^{p^{n+1}}} = z^{ip} g = g$. That is, $h^{p^{n+1}}$ is central. $\qquad\square$

Lemma 4.2.3. *Let R be a ring. If $\eta_1, \ldots, \eta_n \in R$ are central and square-zero, then for any $r_1, \ldots, r_n \in R$, we have*

$$(1 + \eta_1 r_1, \ldots, 1 + \eta_n r_n) = 1 + \eta_1 \cdots \eta_n[r_1, \ldots, r_n].$$

Proof. When $n = 2$, we have

$$(1 + \eta_1 r_1, 1 + \eta_2 r_2) = (1 - \eta_1 r_1)(1 - \eta_2 r_2)(1 + \eta_1 r_1)(1 + \eta_2 r_2) = 1 + \eta_1 \eta_2[r_1, r_2],$$

as required. The remaining cases follow by induction. $\qquad\square$

Recall that for any group G, the lower central series of G is defined via $\gamma_1(G) = G$ and $\gamma_{n+1}(G) = (\gamma_n(G), G)$.

Lemma 4.2.4. *Let R be a ring. Then for any positive integer n, $\gamma_n(\mathcal{U}(R)) \subseteq 1 + R^{(n)}$. In particular, if R is strongly Lie nilpotent, then $\mathcal{U}(R)$ is nilpotent.*

Proof. Our proof is by induction on n. The $n = 1$ case is trivial. Take $u \in \gamma_n(\mathscr{U}(R))$, $v \in \mathscr{U}(R)$. Then

$$(u, v) - 1 = u^{-1}v^{-1}[u, v] = u^{-1}v^{-1}[u - 1, v].$$

By our inductive hypothesis, $u - 1 \in R^{(n)}$, hence $(u, v) \in 1 + R^{(n+1)}$. As $R^{(n+1)}$ is an ideal, we can see that products and inverses of units in $1 + R^{(n+1)}$ also lie in $1 + R^{(n+1)}$. We are done.   □

Proof of Theorem 4.2.1. Suppose that $\mathscr{U}(FG)$ is nilpotent. Surely G is nilpotent hence, if it has an element of order p, then it has one in its centre. Thus, in view of Lemma 4.2.2 and Proposition 1.3.7, we know that G' is a p-group of bounded exponent. We assume that G' is infinite and seek a contradiction.

Let n be the largest positive integer such that $\gamma_n(G)$ is infinite. (Since G is nilpotent, the lower central series eventually reaches 1.) Then $\gamma_{n+1}(G)$ is a finite p-group. Thus, it suffices to show that $G'/\gamma_{n+1}(G)$ is finite. But by Lemma 1.2.18, $\mathscr{U}(F(G/\gamma_{n+1}(G)))$ is nilpotent. Thus, we can factor out $\gamma_{n+1}(G)$ and assume that $\gamma_n(G)$ is an infinite central p-group of bounded exponent. By the Prüfer–Baer theorem, G contains a central subgroup of the form $A = \prod_{i=1}^{\infty} A_i$, where each A_i is a nontrivial cyclic p-group.

Let X be a transversal of A in G. Suppose that $\mathscr{U}(FG)$ satisfies $(x_1, \ldots, x_m) = 1$, and take any $\alpha_1, \ldots, \alpha_m \in FG$. We may choose a positive integer k so that $[\alpha_1, \ldots, \alpha_m] = \sum_j \beta_j h_j$, where each $\beta_j \in F(A_1 \times \cdots \times A_k)$ and $h_j \in X$. For each i, $1 \le i \le m$, let $\eta_i = \hat{A}_{k+i}$. Each η_i is central and square-zero; hence, by Lemma 4.2.3,

$$\begin{aligned}
1 &= (1 + \eta_1\alpha_1, \ldots, 1 + \eta_m\alpha_m) \\
&= 1 + \eta_1 \cdots \eta_m[\alpha_1, \ldots, \alpha_m] \\
&= 1 + \eta_1 \cdots \eta_m \sum_j \beta_j h_j.
\end{aligned}$$

As the h_j lie in distinct cosets of A in G, this means that $\eta_1 \cdots \eta_m \beta_j = 0$ for all j. But the product of the A_i is direct. Thus, as no $\eta_i = 0$, we must have $\beta_j = 0$ for all j, hence $[\alpha_1, \ldots, \alpha_m] = 0$. That is, FG is Lie nilpotent. By Theorem 3.1.1, G' is a finite p-group.

For the sufficiency we combine Lemmas 4.2.4 and 3.3.4.   □

Thus, it remains to consider semiprime group rings. If G is torsion, the result follows in a straightforward manner. The following lemma will be useful later, too.

Lemma 4.2.5. *Let R be a commutative ring. Fix $r \in R$ and a positive integer n. Then, working in $GL_2(R)$, we have*

$$\left(\begin{pmatrix} 1 & r \\ 0 & 1 \end{pmatrix}, \underbrace{\begin{pmatrix} 1 & 0 \\ r & 1 \end{pmatrix}, \ldots, \begin{pmatrix} 1 & 0 \\ r & 1 \end{pmatrix}}_{n \text{ times}} \right) = \begin{pmatrix} a_{11} & r^{2^{n+1}-1} \\ a_{21} & a_{22} \end{pmatrix},$$

for some $a_{11}, a_{21}, a_{22} \in R$.

Proof. An easy computation reveals that for any matrix

$$\begin{pmatrix} a & b \\ c & d \end{pmatrix}$$

of determinant 1, we have

$$\left(\begin{pmatrix} a & b \\ c & d \end{pmatrix}, \begin{pmatrix} 1 & 0 \\ r & 1 \end{pmatrix} \right) = \begin{pmatrix} a_{11} & b^2 r \\ a_{21} & a_{22} \end{pmatrix}$$

for some $a_{11}, a_{21}, a_{22} \in R$. The lemma now follows by induction. $\qquad\square$

In particular, taking $r = 1$, we can see that $GL_2(F)$ is not nilpotent for any field F. The result of Bateman and Coleman [5] for finite groups now extends to torsion groups in a natural way.

Proposition 4.2.6. *Let char $F = p \geq 0$ and let G be a torsion group. Then $\mathcal{U}(FG)$ is nilpotent if and only if G is nilpotent and p-abelian.*

Proof. If $p > 0$ and G has a p-element, then Theorem 4.2.1 gives us the result. Thus, we may assume that FG is semiprime. We must show that if $\mathcal{U}(FG)$ is nilpotent, then G is abelian. Thus, it suffices to consider finitely generated groups G. Now, a torsion nilpotent group is locally finite, hence we may take G to be finite. But then FG is semisimple, and

$$FG \cong M_{n_1}(D_1) \oplus \cdots \oplus M_{n_k}(D_k),$$

where the n_i are positive integers and the D_i are division algebras over F. Thus, each $GL_{n_i}(D_i)$ is nilpotent. By Proposition 1.2.2, D_i is a field and, as we have just seen, each $n_i = 1$. Thus, FG is commutative, and we are done. $\qquad\square$

As a nice consequence, we have

Corollary 4.2.7. *Let F be any field and G a torsion group. Then $\mathcal{U}(FG)$ is nilpotent if and only if FG is Lie nilpotent.*

Proof. Combine the proposition with Theorem 3.1.1. $\qquad\square$

Finally, suppose that FG is semiprime and G is not torsion. There is an exceptional case involving fields of prime order. We will bypass it and assume that F is infinite.

Let R be a ring and F a field contained in R (not necessarily as a unital subring). Suppose $u \in \mathcal{U}(R)$ and $\langle u \rangle$ normalizes F. Then we can define a $\mathbb{Z}\langle u \rangle$-module action on $\mathcal{U}(F)$ as follows. Let $\alpha = \sum_{i \in \mathbb{Z}} \alpha_i u^i \in \mathbb{Z}\langle u \rangle$. Then for any $\lambda \in \mathcal{U}(F)$, we let $\lambda^\alpha = \prod_{i \in \mathbb{Z}} (\lambda^{\alpha_i})^{u^i}$.

In particular, we can define $(\lambda, u) = \lambda^{-1} \lambda^u = \lambda^{u-1}$. Thus, for any positive integer n,

$$(\lambda, \underbrace{u, \ldots, u}_{n \text{ times}}) = \lambda^{(u-1)^n}.$$

Lemma 4.2.8. *Using the notation above, suppose that F is infinite and*

$$(\lambda, \underbrace{u, \ldots, u}_{n \text{ times}}) = 1$$

for all $0 \neq \lambda \in F$. If, for some positive integer m, u^m centralizes F, then u centralizes F.

Proof. As we observed above, $\lambda^{(u-1)^n} = 1$ for all $0 \neq \lambda \in F$. But also, $\lambda^{u^m-1} = \lambda^{u^m}\lambda^{-1} = 1$. As polynomials in $\mathbb{Q}[x]$, we see that the greatest common divisor of $(x-1)^n$ and $x^m - 1$ is $x - 1$. Thus, there exist $f(x), g(x) \in \mathbb{Z}[x]$ and a positive integer k such that $f(x)(x-1)^n + g(x)(x^m - 1) = k(x-1)$. In particular, $f(u)(u-1)^n + g(u)(u^m - 1) = k(u-1)$. Thus, $\lambda^{k(u-1)} = 1$, hence $(\lambda^k)^u = \lambda^k$ for all $\lambda \in F$.

As F contains only finitely many kth roots of unity, F^k is infinite. Fix any $0 \neq \lambda \in F$. Then $\lambda^k = (\lambda^k)^u = (\lambda^u)^k$, hence $\lambda^u = \lambda\xi_1$, where $\xi_1 \in F$ is a kth root of unity. Also, if $\mu \in F^k$, then by the same argument, $(\lambda + \mu)^u = (\lambda + \mu)\xi_2$, where $\xi_2 \in F$ is a kth root of unity. As there are infinitely many such μ and only finitely many kth roots of unity, we may fix a ξ_2 for which there are infinitely many such μ. Of course, $(\lambda + \mu)^u = \lambda^u + \mu^u = \lambda\xi_1 + \mu$. Therefore, $(\lambda + \mu)\xi_2 = \lambda\xi_1 + \mu$, hence

$$\lambda(\xi_2 - \xi_1) = \mu(1 - \xi_2).$$

But the left-hand side of this last expression is fixed; thus, choosing distinct values μ_1 and μ_2 for μ, we get $(\mu_1 - \mu_2)(1 - \xi_2) = 0$. That is, $\xi_2 = 1$. Since $\lambda \neq 0$, $\xi_1 = 1$ as well. Thus, λ commutes with u, as required. □

The main result of Fisher et al. [30] and Khripta [55] is the following. Recall that a Mersenne prime is a prime of the form $2^q - 1$ for some prime q.

Theorem 4.2.9. *Let F be a field of characteristic $p \geq 0$ and G a group such that, if $p > 0$, then G has no p-elements. Then $\mathscr{U}(FG)$ is nilpotent if and only if G is nilpotent, the torsion elements of G form an abelian (normal) subgroup T and either*

1. *T is central in G, or*
2. *$|F| = p$ is a Mersenne prime, $T^{p^2-1} = 1$, and for every $h \in T$, $g \in G$, we have $h^g = h$ or h^p.*

Proof. If F is finite, then we refer the reader to the proof of [94, Theorem VI.3.6]. We assume that F is infinite. Let $\mathscr{U}(FG)$ be nilpotent. Clearly, G is nilpotent, hence T is a subgroup. By Proposition 4.2.6, T is abelian. It suffices to show that every element of infinite order commutes with every torsion element; that is, we may assume that G is finitely generated. Now, every subgroup of a finitely generated nilpotent group is finitely generated (see, for instance, [94, Corollary I.3.10]). Thus, T is finite and FT is semisimple.

Write

$$FT = FTe_1 \oplus \cdots \oplus FTe_k,$$

where each e_i is a primitive idempotent of FT. Then each FTe_i is a field. If $G = T$, there is nothing to do, so assume that G is not torsion. Then by Theorem 1.4.9, e_i is central in FG. Thus, if $g \in G$ has infinite order, then $\langle g \rangle$ normalizes FTe_i. Suppose that $\mathscr{U}(FG)$ satisfies $(x_1, \ldots, x_n) = 1$. Then, working in $\mathscr{U}(FGe_i)$, we see that for any $\lambda \in \mathscr{U}(FTe_i)$, we have

$$(\lambda, \underbrace{ge_i, \ldots, ge_i}_{n-1 \text{ times}}) = e_i.$$

Furthermore, as T is finite, its automorphism group is finite, hence g^m centralizes FTe_i for some m. By Lemma 4.2.8, g centralizes FTe_i for all i, hence g centralizes FT, and T is a central subgroup.

Conversely, suppose that G satisfies $(x_1, \ldots, x_n) = 1$ and T is central. We claim that $\mathscr{U}(FG)$ satisfies $(x_1, \ldots, x_n) = 1$. Thus, it suffices to assume that G is finitely generated; hence, as we saw above, that T is finite. Let $e_1, \ldots, e_k$ be the primitive idempotents of FT (which, of course, are central in FG). By Remark 1.4.10, every unit of FGe_i is of the form λh, with $\lambda \in FTe_i$, $h \in G$. As FTe_i is central in FGe_i, we get $(\lambda_1 h_1, \ldots, \lambda_n h_n) = (h_1, \ldots, h_n)e_i$ for all $\lambda_j \in \mathscr{U}(FTe_i)$, $h_j \in G$. But $(h_1, \ldots, h_n) = 1$, hence $(\lambda_1 h_1, \ldots, \lambda_n h_n) = e_i$. Performing this for every e_i, we see that $\mathscr{U}(FG)$ satisfies $(x_1, \ldots, x_n) = 1$, as required. $\qquad \square$

4.3 Group Rings of Locally Finite Groups

We now begin our discussion of the author's work in [67] classifying the torsion groups G such that $\mathscr{U}^+(FG)$ is nilpotent. If F has characteristic zero, there is nothing left to do. The results of Chapter 2 showed that if $\mathscr{U}^+(FG)$ satisfies a group identity, then the symmetric elements commute. Thus, we will let G be a torsion group and F a field of characteristic $p > 2$.

In this section we consider the problem for locally finite groups, modulo the prime group ring case, with which we will dispense in the next section. The prime group rings were classified by Connell. The following is [82, Theorem 4.2.10].

Proposition 4.3.1. *Let F be a field and G a group. Then the following are equivalent:*

(i) FG is prime;
(ii) G has no nonidentity finite normal subgroups;
(iii) $\phi(G)$ is torsion-free abelian.

Our goal here is to prove the next two propositions.

Proposition 4.3.2. *Let F be a field of characteristic $p > 2$ and G a locally finite group not containing Q_8. If $\mathscr{U}^+(FG)$ is nilpotent, then $G = H \times P$ where H is an* abelian p'-group and P is a p-group. If FP is not a prime ring, then P is nilpotent and P' is finite.

Proposition 4.3.3. Let F be a field of characteristic $p > 2$ and G a locally finite group containing Q_8. If $\mathscr{U}^+(FG)$ is nilpotent, then $G \simeq Q_8 \times E \times P$, where E is an elementary abelian 2-group and P is a p-group. If FP is not a prime ring, then P is finite.

We begin with

Lemma 4.3.4. Suppose char $F \neq 2$ and G is a finite group. If $\mathscr{U}^+(FG)$ is nilpotent, then the 2-elements of G form a (normal) subgroup.

Proof. Our proof is by induction on $|G|$. If G has no 2-elements, there is nothing to do. Otherwise, let S be the set of elements of order 2 in G. Then S consists of symmetric units hence, by Lemma 4.1.2, the subgroup N generated by S is nilpotent. As it is generated by 2-elements, N is a 2-group. By Lemma 2.4.6, $\mathscr{U}^+(F(G/N))$ is nilpotent. Thus, our inductive hypothesis tells us that the 2-elements of G/N form a group. Since N is a 2-group, we are done.   □

Lemma 4.3.5. Suppose char $F = p > 2$ and G is a group with $\mathscr{U}^+(FG)$ nilpotent. Let g be a p-element of G. If h is a p'-element of G whose order is 2 or odd, then $gh = hg$.

Proof. Lemma 4.1.2 tells us that the subgroup of $\mathscr{U}(FG)$ generated by the symmetric units is nilpotent. Thus, any two torsion symmetric units of relatively prime order commute. Let us say that $o(g) = p^m$ and $o(h) = 2$. Then we notice that $(\frac{g+g^{-1}}{2})^{p^m} = 1$, hence $[g + g^{-1}, h] = 1$. That is,

$$gh + g^{-1}h = hg + hg^{-1},$$

hence $gh = hg$ (as required) or $gh = hg^{-1}$. But in the latter case, $\langle g, h \rangle$ is dihedral of order $2p^m$. However, the 2-elements of this group do not form a subgroup, contradicting the previous lemma.

On the other hand, suppose that $o(h) = k$, where neither 2 nor p divides k. Choosing l so that $p^l \equiv 1 \pmod{k}$, we get $(h + h^{-1})^{p^l} = h + h^{-1}$. But also,

$$(h + h^{-1})h \left(\frac{1 - h^2 + h^4 - \cdots + h^{2(k-1)}}{2} \right) = \frac{1 + h^{2k}}{2} = 1.$$

Thus, $h + h^{-1}$ is a symmetric unit of order dividing $p^l - 1$; that is, it has p'-order. Hence, $[g + g^{-1}, h + h^{-1}] = 0$.

By Lemma 3.2.2, $gh \in \{hg, h^{-1}g, hg^{-1}, h^{-1}g^{-1}\}$. Suppose that $gh \neq hg$. If $gh = h^{-1}g$, then $h^g = h^{-1}$, hence g^2 commutes with h. As g has odd order, g commutes with h. Similarly, if $gh = hg^{-1}$. Finally, if $gh = h^{-1}g^{-1}$, then $(gh)^2 = 1$; hence, by the first part of the proof, gh commutes with g. Thus, $gh = hg$.   □

We can now extend Lemma 4.3.4 to all p'-elements.

Lemma 4.3.6. *Let char $F = p > 2$ and let G be a finite group. If $\mathcal{U}^+(FG)$ is nilpotent, then the p'-elements of G form a (normal) subgroup.*

Proof. Let P denote the set of p-elements of G and let $H = C_G(P)$. Note that H is a subgroup of G containing every element of odd p'-order, by the preceding lemma. Now, $H \cap P$ is central in H, hence it is a subgroup, and $H/(H \cap P)$ is a p'-group. By the Schur–Zassenhaus theorem, $H = (H \cap P) \rtimes K$ for some p'-subgroup K. But since $H \cap P$ is central, we have $H = (H \cap P) \times K$. Thus, K contains every element of odd p'-order in H, hence in G. By Lemma 4.3.4, the 2-elements of G form a normal subgroup L. Thus, KL is a p'-group containing every 2-element and every odd p'-element. We are done. □

Let us now restrict the form of the p'-subgroup mentioned in the last lemma. We will need

Lemma 4.3.7. *Let F be a field of characteristic different from 2 and n a positive integer. Let $*$ be an involution on $M_n(F)$. If the symmetric units in $M_n(F)$ generate a nilpotent group, then all symmetric matrices in $M_n(F)$ are central.*

Proof. If $n = 1$, there is nothing to do, so let us assume that $n \geq 2$. Write H for the subgroup of $GL_n(F)$ generated by the symmetric units. In view of Proposition 2.1.4, there are two cases to consider. Suppose, first of all, that $*$ is of transpose type. Then let U be the matrix mentioned in Proposition 2.1.4, and let $^-$ be the involution on F. Letting F' be the prime subfield of F, we see that F' is fixed elementwise by $^-$.

Denoting the usual transpose involution by t, suppose that $M \in GL_n(F')$ is a matrix such that $M^t = M$. Then we note that $(MU)^* = U^{-1}U^t M^t U = MU$ (since the entries of U are also fixed by $^-$). Of course, $U^* = U$ as well. Thus, H contains $MUU^{-1} = M$ for all such M. It will now suffice to show that H is not nilpotent when $n = 2$, as the cases for larger n will follow immediately.

But we now know that H contains

$$\begin{pmatrix} 0 & 1 \\ 1 & 0 \end{pmatrix}, \ \begin{pmatrix} 0 & 1 \\ 1 & 1 \end{pmatrix} \ \text{and} \ \begin{pmatrix} 1 & 1 \\ 1 & 0 \end{pmatrix}.$$

Thus, H contains

$$\begin{pmatrix} 0 & 1 \\ 1 & 0 \end{pmatrix} \begin{pmatrix} 0 & 1 \\ 1 & 1 \end{pmatrix} = \begin{pmatrix} 1 & 1 \\ 0 & 1 \end{pmatrix}$$

and

$$\begin{pmatrix} 0 & 1 \\ 1 & 0 \end{pmatrix} \begin{pmatrix} 1 & 1 \\ 1 & 0 \end{pmatrix} = \begin{pmatrix} 1 & 0 \\ 1 & 1 \end{pmatrix}.$$

But by Lemma 4.2.5, these matrices do not generate a nilpotent group. This case is complete.

So, suppose that $*$ is the canonical symplectic involution. Then $n = 2m$. If $m = 1$, then the symmetric matrices are simply the scalar multiples of the identity matrix, which, of course, are central. Thus, let us assume that $m \geq 2$. Take any $A \in GL_m(F)$. Then

$$\begin{pmatrix} A & 0 \\ 0 & A^t \end{pmatrix}$$

is a symmetric unit. Thus, if H is nilpotent, then so is $GL_m(F)$. But by Lemma 4.2.5, this is not the case. We are done.   □

Lemma 4.3.8. *Let char $F = p > 2$ and let G be a finite p'-group. If $\mathscr{U}^+(FG)$ is nilpotent, then G is abelian or a Hamiltonian 2-group.*

Proof. It will suffice to assume that $F = \mathbb{Z}_p$. As FG is semisimple, write

$$FG = FGe_1 \oplus FGe_2 \oplus \cdots \oplus FGe_k,$$

where each e_i is a primitive central idempotent. Let us say that each $FGe_i \cong M_{n_i}(D_i)$, where D_i is a finite division ring, hence a field.

Suppose that e_1 is not symmetric. Then e_1^* is also a primitive central idempotent, say e_2. Take any $\alpha e_1 \in \mathscr{U}(FGe_1)$. Then

$$\alpha e_1 + \alpha^* e_2 + e_3 + \cdots + e_k \in \mathscr{U}^+(FG).$$

Thus, under the natural projection $FG \to FGe_1$, the symmetric units map onto $GL_{n_1}(D_1)$. By Lemma 4.2.5, $GL_{n_1}(D_1)$ is not nilpotent if $n_1 \geq 2$. Thus, $n_1 = 1$.

On the other hand, suppose that $e_1^* = e_1$. Then $*$ induces an involution of $M_{n_1}(D_1)$. Furthermore, if αe_1 is a symmetric unit of FGe_1, then $\alpha e_1 + e_2 + \cdots + e_k \in \mathscr{U}^+(FG)$. Thus the symmetric units of FG map onto the symmetric units of FGe_1. In particular, the symmetric units of $GL_{n_1}(D_1)$ generate a nilpotent group. By Lemma 4.3.7, it follows that the symmetric elements of $M_{n_1}(D_1)$ commute.

Thus, in any case, the projections of $(FG)^+$ onto each of the Wedderburn components are commutative. Therefore, $(FG)^+$ is commutative. In particular, it is Lie nilpotent, and by Theorems 3.3.1 and 3.3.6, G is abelian or a Hamiltonian 2-group, since G has no p-elements. We are done.   □

What, then, of the p-elements of G?

Lemma 4.3.9. *Let char $F = p > 2$ and let G be a finite group. If $\mathscr{U}^+(FG)$ is nilpotent, then the p-elements of G form a (normal) subgroup.*

Proof. Our proof is by induction on $|G|$. By Lemma 4.3.6, the p'-elements form a normal subgroup H of G. Hence, by the Schur–Zassenhaus theorem, $G = H \rtimes K$ for some p-subgroup K. Suppose that G has no 2-elements. Since by Lemma 4.3.5, odd p'-elements commute with p-elements, we have $G = H \times K$, and we are done.

Otherwise, let N be the subgroup of H generated by the elements of order 2. Lemma 4.3.5 tells us that N centralizes K. Furthermore, by the last lemma, H is abelian or a Hamiltonian 2-group. Either way, elements of order 2 are central in H. Thus, N is central in G. By Lemma 2.4.6, $\mathscr{U}^+(F(G/N))$ is nilpotent. Therefore, our inductive hypothesis tells us that the p-elements of G/N form a normal subgroup,

L/N. Applying the Schur–Zassenhaus theorem to L, we get $L = N \rtimes M$, for some p-subgroup M. Since N is central, this means that $L = N \times M$. But L contains every p-element of G, hence M consists of all of the p-elements, and the proof is complete.
□

We have now completed the proofs of Propositions 4.3.2 and 4.3.3 for finite groups G. If G is locally finite, then by considering finite subgroups, we obtain the first part of each of these propositions, namely, that $G = H \times P$ where P is a p-group and H is a p'-group, such that H is abelian or a Hamiltonian 2-group. Let us narrow down the possibilities for P.

Lemma 4.3.10. *Let G be a group and F a field of characteristic $p > 0$. Suppose that G contains an infinite direct product of normal subgroups $A = A_1 \times A_2 \times \cdots$, each of which is a nontrivial finite p-group. If $\mathscr{U}^+(FG)$ satisfies $(x_1, \ldots, x_m) = 1$, then $(FG)^+$ satisfies $[x_1, \ldots, x_m] = 0$.*

Proof. Let X be a transversal of A in G, and take any $\alpha_1, \ldots, \alpha_m \in (FG)^+$. Choose a positive integer k so that $[\alpha_1, \ldots, \alpha_m] = \sum_j \beta_j h_j$, where each $\beta_j \in F(A_1 \times \cdots \times A_k)$ and $h_j \in X$. For each i, $1 \le i \le m$, let $\eta_i = \hat{A}_{k+i}$. Each η_i is central, symmetric and square-zero. Therefore, $1 + \eta_i \alpha_i \in \mathscr{U}^+(FG)$. Hence, by Lemma 4.2.3,

$$1 = (1 + \eta_1 \alpha_1, \ldots, 1 + \eta_m \alpha_m)$$
$$= 1 + \eta_1 \cdots \eta_m [\alpha_1, \ldots, \alpha_m]$$
$$= 1 + \eta_1 \cdots \eta_m \sum_j \beta_j h_j.$$

As the h_j lie in distinct cosets of A in G, $\eta_1 \cdots \eta_m \beta_j = 0$ for all j. But the product of the A_i is direct. Thus, as no $\eta_i = 0$, we must have $\beta_j = 0$ for all j; hence $[\alpha_1, \ldots, \alpha_m] = 0$, as required.
□

Lemma 4.3.11. *Let char $F = p > 2$ and let G be a nilpotent p-group with G' of bounded exponent. If $\mathscr{U}^+(FG)$ is nilpotent, then G' is finite.*

Proof. Suppose that G' is infinite. Then, since G is nilpotent, there is a largest positive integer k such that $\gamma_k(G)$ is infinite. By Lemma 2.3.6, $\mathscr{U}^+(F(G/\gamma_{k+1}(G)))$ is nilpotent. It will suffice to show that $G'/\gamma_{k+1}(G)$ is finite. Thus, we factor out $\gamma_{k+1}(G)$ and assume that $\gamma_k(G)$ is an infinite central group. It also has bounded exponent. Thus, by the Prüfer–Baer theorem, $\gamma_k(G)$ is an infinite direct product of finite p-groups. The previous lemma says that $(FG)^+$ is Lie nilpotent. Thus, by Theorem 3.3.1, G' is finite.
□

We also need the following well-known group-theoretic result.

Lemma 4.3.12. *If G is a p-group and G' is finite, then G is nilpotent.*

Proof. Our proof is by induction on $|G'|$. If $G' = 1$, then there is nothing to do, so assume that this is not the case. As G' is surely nilpotent, let k be the largest positive integer such that $\gamma_k(G')$ is nontrivial. Then $\gamma_k(G')$ is abelian. Since $\gamma_k(G')$ is finite and normal, it is contained in $\phi(G)$. Thus, $G/C_G(\gamma_k(G'))$ is finite. By Lemma 3.2.6, there exists an m so that

$$(\gamma_k(G'), \underbrace{G, \ldots, G}_{m \text{ times}}) = 1.$$

But $|(G/\gamma_k(G'))'| = |G'/\gamma_k(G')| < |G'|$. Thus, by our inductive hypothesis, $G/\gamma_k(G')$ is nilpotent. Hence, for some n,

$$\underbrace{(G, \ldots, G)}_{n \text{ times}} \leq \gamma_k(G')$$

and therefore $(G, \ldots, G) = 1$, as required.   □

We can now prove Proposition 4.3.2. The proof includes a simplification from Lee and Spinelli [74] which will also be useful when we discuss the bounded Engel property.

Proof of Proposition 4.3.2. Considering finitely generated (hence finite) subgroups, we see from Lemmas 4.3.6 and 4.3.9 that $G = H \times P$, where H is a p'-group and P is a p-group. Since G does not contain Q_8, Lemma 4.3.8 tells us that H is abelian. Thus, we may as well assume that G is a locally finite p-group.

Since we are assuming that FG is not prime, Proposition 4.3.1 says that G has a nontrivial finite normal subgroup N. It will suffice to show that $(G/N)'$ is finite; indeed, then $G'N/N$ is finite, hence G' is finite and by the last lemma, it follows immediately that G is nilpotent.

Let $\eta = \hat{N}$, and take $\alpha \in (FG)^+$, $\beta \in \mathscr{U}^+(FG)$. Suppose that $\mathscr{U}^+(FG)$ satisfies $(x_1, \ldots, x_{p^m+1}) = 1$. Since $1 + \eta\alpha \in \mathscr{U}^+(FG)$, we see from Lemma 4.1.1 that

$$1 = (1 + \eta\alpha, \underbrace{\beta, \ldots, \beta}_{p^m \text{ times}}) = 1 + \eta(\alpha^{\beta^{p^m}} - \alpha).$$

Thus, $\eta(\alpha^{\beta^{p^m}} - \alpha) = 0$. By Lemma 1.3.8, $\alpha^{\beta^{p^m}} - \alpha \in \Delta(G, N)$. Thus, working in $F\bar{G} = F(G/N)$, we have

$$[\bar\alpha, \bar\beta^{p^m}] = 0.$$

Now, G is a locally finite p-group. Letting $\varepsilon : FG \to F$ be the augmentation map, we notice that for any $\beta' \in FG$, we have $\beta' - \varepsilon(\beta') \in \Delta(G)$, which is nil, by Lemma 1.1.1. Thus, for a suitable r, $(\beta')^{p^r} = (\varepsilon(\beta'))^{p^r}$. That is, if β' does not have augmentation 0, then β' has a power which is a nonzero field element; hence, β' is a unit. In particular, then, for any $\beta' \in (FG)^+$, either $\beta' \in \mathscr{U}^+(FG)$ or $\beta' + 1 \in \mathscr{U}^+(FG)$. In the latter case, $[\bar\alpha, \bar{1} + (\bar{\beta'})^{p^m}] = 0$ implies that $[\bar\alpha, (\bar{\beta'})^{p^m}] = 0$. Thus, we have

$$0 = [\bar\alpha, \bar\beta^{p^m}] = [\bar\alpha, \underbrace{\bar\beta, \ldots, \bar\beta}_{p^m \text{ times}}],$$

for all $\bar\alpha, \bar\beta \in (F\bar{G})^+$. That is, $(F\bar{G})^+$ is Lie p^m-Engel.

Thus, by Lemma 3.2.5, $\bar{G}/\zeta(\bar{G})$ is a p-group of bounded exponent. Also, by Theorem 3.2.8, $\bar{G}$ is nilpotent. By Proposition 1.3.7, $(\bar{G})'$ has bounded exponent. Thus, by Lemma 4.3.11, $(\bar{G})'$ is finite. The proof is complete.   □

Conversely, of course, if G is nilpotent and p-abelian, then we know that $\mathscr{U}(FG)$ is nilpotent.

Let us now consider groups containing the quaternions.

Lemma 4.3.13. *Let* char $F = p > 2$, *and let* $G = Q_8 \times \langle c \rangle$, *where c is a p-element. If* $\mathscr{U}^+(FG)$ *satisfies* $(x_1, \ldots, x_n) = 1$, *then* $o(c) \leq 2^{n+1} - 2$.

Proof. In the proof of Lemma 2.3.7 we saw that there is a homomorphism $\theta : FG \rightarrow M_2(F\langle c \rangle)$ and we found $\alpha, \beta \in (FG)^+$ such that $\alpha^2 = \beta^2 = 0$; and, furthermore,

$$\theta(\alpha) = \begin{pmatrix} 0 & c^{-1} - c \\ 0 & 0 \end{pmatrix} \text{ and } \theta(\beta) = \begin{pmatrix} 0 & 0 \\ c^{-1} - c & 0 \end{pmatrix}.$$

(While F was assumed to be infinite, that fact was not needed for this construction.)

Then $1 + \alpha, 1 + \beta \in \mathscr{U}^+(FG)$ and

$$\theta(1 + \alpha) = \begin{pmatrix} 1 & c^{-1} - c \\ 0 & 1 \end{pmatrix} \text{ and } \theta(1 + \beta) = \begin{pmatrix} 1 & 0 \\ c^{-1} - c & 1 \end{pmatrix}.$$

Since

$$(1 + \alpha, \underbrace{1 + \beta, \ldots, 1 + \beta}_{n-1 \text{ times}}) = 1,$$

we must have

$$\begin{pmatrix} 1 & 0 \\ 0 & 1 \end{pmatrix} = \left(\begin{pmatrix} 1 & c^{-1} - c \\ 0 & 1 \end{pmatrix}, \underbrace{\begin{pmatrix} 1 & 0 \\ c^{-1} - c & 1 \end{pmatrix}, \ldots, \begin{pmatrix} 1 & 0 \\ c^{-1} - c & 1 \end{pmatrix}}_{n-1 \text{ times}} \right)$$

$$= \begin{pmatrix} \alpha_{11} & (c^{-1} - c)^{2^n - 1} \\ \alpha_{21} & \alpha_{22} \end{pmatrix},$$

for some $\alpha_{11}, \alpha_{21}, \alpha_{22} \in F\langle c \rangle$, by Lemma 4.2.5. That is, $(c^{-1} - c)^{2^n - 1} = 0$, hence

$$0 = c^{2^n - 1}(c^{-1} - c)^{2^n - 1} = (1 - c^2)^{2^n - 1}.$$

Thus, the leading term, $-c^{2^{n+1} - 2}$, must cancel with some lower term, hence $o(c) \leq 2^{n+1} - 2$, as required.   □

Finally, we have the

Proof of Proposition 4.3.3. In view of Lemmas 4.3.6 and 4.3.9, we have $G \simeq H \times P$, where H is a p'-group and P is a p-group. By Lemma 4.3.8, H is a Hamiltonian 2-group. Furthermore, in view of Lemma 4.3.13, P has bounded exponent. Suppose that P is infinite and FP is not prime.

By Proposition 4.3.2, P' is finite. Now, Lemma 2.3.6 tells us that $\mathcal{U}^+(F(G/P'))$ is nilpotent. But $G/P' \simeq Q_8 \times E \times (P/P')$, where E is an elementary abelian 2-group and P/P' is an infinite central p-group of bounded exponent. Thus, P/P' is an infinite direct product of cyclic groups. By Lemma 4.3.10, $(F(G/P'))^+$ is Lie nilpotent. But this contradicts Theorem 3.3.6. We are done.   □

In fact, if G is the direct product of a Hamiltonian 2-group and a finite p-group, then $\mathcal{U}^+(FG)$ is nilpotent, but we will postpone the proof of this.

4.4 Prime Group Rings

Our next step is to eliminate the case in which FG is a prime ring. We will prove

Proposition 4.4.1. *Let F be a field of characteristic different from 2 and G a torsion group such that FG is prime. If $\mathcal{U}^+(FG)$ satisfies a group identity, then G is the trivial group.*

If F is infinite, then this follows immediately from the classification in Chapter 2, so our concern is the finite field case.

In order to prove this result, we need to know something about generalized *-polynomial identities. These were discussed extensively by Rowen in [90] and [91]. If R is an F-algebra with involution, then a generalized *-polynomial identity (*-GPI) $f(x_1, x_1^*, \ldots, x_n, x_n^*)$ is a sum of terms of the form

$$r_0 x_{i_1}^{\varepsilon_1} r_1 x_{i_2}^{\varepsilon_2} r_2 \cdots r_{k-1} x_{i_k}^{\varepsilon_k} r_k,$$

where each $r_i \in R$, each ε_i is equal to 1 or *, and k is an integer, such that $f(a_1, a_1^*, \ldots, a_n, a_n^*) = 0$ for all $a_1, \ldots, a_n \in R$.

In particular, a multilinear *-GPI has the form

$$f(x_1, x_1^*, \ldots, x_n, x_n^*) = \sum_{\sigma \in S_n} \sum_{\varepsilon_i \in \{1, *\}} f^{(\sigma, \varepsilon_1, \ldots, \varepsilon_n)}(x_1, x_1^*, \ldots, x_n, x_n^*),$$

where each $f^{(\sigma, \varepsilon_1, \ldots, \varepsilon_n)}$ is a sum of terms of the form

$$r_0 x_{\sigma(1)}^{\varepsilon_1} r_1 x_{\sigma(2)}^{\varepsilon_2} r_2 \cdots r_{n-1} x_{\sigma(n)}^{\varepsilon_n} r_n,$$

with each $r_i \in R$ and such that $f(a_1, a_1^*, \ldots, a_n, a_n^*) = 0$ for all $a_1, \ldots, a_n \in R$. For each $\sigma \in S_n$, $\varepsilon_i \in \{1, *\}$, let $g^{(\sigma, \varepsilon_1, \ldots, \varepsilon_n)}(x_1, x_2, \ldots, x_{2n})$ be the expression obtained by replacing each x_i^* with x_{n+i} in $f^{(\sigma, \varepsilon_1, \ldots, \varepsilon_n)}$. We say that f is a nondegenerate multilinear *-GPI if there exist $\sigma \in S_n$, $\varepsilon_i \in \{1, *\}$ such that $g^{(\sigma, \varepsilon_1, \ldots, \varepsilon_n)}$ is not a GPI for R.

We note that the linearization process described in Section 1.2 extends easily to this situation. If there exists an i such that neither x_i nor x_i^* appears in a term in the sum described above, then we may substitute $x_i = x_i^* = 0$ and obtain a *-GPI in fewer variables. Thus, let us assume that either x_i or x_i^* appears in every such term for all i. Choose an i so that either x_i or x_i^* appears more than once, or both occur, in some term. For the sake of simplicity, say $i = n$. Then R also satisfies

$$f(x_1, x_1^*, \ldots, x_n + x_{n+1}, x_n^* + x_{n+1}^*) - f(x_1, x_1^*, \ldots, x_n, x_n^*)$$
$$- f(x_1, x_1^*, \ldots, x_{n-1}, x_{n-1}^*, x_{n+1}, x_{n+1}^*).$$

Repeating this procedure, we will eventually obtain a multilinear $*$-GPI. Of course, it does not necessarily follow that the $*$-GPI so obtained will be nondegenerate. If it is nondegenerate, then we will have nothing more to do, due to the following classical result of Rowen.

Proposition 4.4.2. *Let F be a field and R a prime F-algebra. If R satisfies a nondegenerate multilinear $*$-GPI, then R satisfies a nondegenerate multilinear GPI.*

Proof. See [90, Theorem 9]. ☐

From this, we immediately get

Lemma 4.4.3. *Let F be a field and G a torsion group. If FG is prime and satisfies a nondegenerate multilinear $*$-GPI, then G is the trivial group.*

Proof. By Propositions 4.4.2 and 1.2.15, $(G : \phi(G)) < \infty$. But, by Proposition 4.3.1, $\phi(G) = 1$. Thus, G is finite, hence $G = \phi(G) = 1$. ☐

In fact, for simple sorts of generalized $*$-polynomials, it is easy (and useful) to determine precisely what we will obtain at the end of the linearization process. Indeed, it is a straightforward exercise to check that if

$$f(x_1, x_1^*) = r_0 x_1^{\varepsilon_1} r_1 x_1^{\varepsilon_2} r_2 \cdots r_{n-1} x_1^{\varepsilon_n} r_n,$$

with each $r_i \in R$ and $\varepsilon_i \in \{1, *\}$, then the result of the linearization will be

$$g(x_1, x_1^*, \ldots, x_n, x_n^*) = \sum_{\sigma \in S_n} r_0 x_{\sigma(1)}^{\varepsilon_1} r_1 x_{\sigma(2)}^{\varepsilon_2} r_2 \cdots r_{n-1} x_{\sigma(n)}^{\varepsilon_n} r_n.$$

We will obtain helpful information regardless of whether this turns out to be nondegenerate.

Let us begin with this result.

Lemma 4.4.4. *Let G be torsion and let FG be prime. Fix any $\alpha, \beta \in FG$ and let n be a positive integer. If $(\alpha \gamma \beta \gamma^*)^n = 0$ for all $\gamma \in FG$, then $\alpha = 0$ or $\beta = 0$.*

Proof. We know that FG satisfies the $*$-GPI

$$(\alpha x_1 \beta x_1^*)^n = 0.$$

Linearizing this expression as we did above, we obtain the multilinear $*$-GPI

$$\sum_{\sigma \in S_{2n}} \alpha x_{\sigma(1)} \beta x^*_{\sigma(2)} \cdots \alpha x_{\sigma(2n-1)} \beta x^*_{\sigma(2n)} = 0.$$

If this is nondegenerate, then by the previous lemma, $G = 1$, and taking $\gamma = 1$ gives the result. So assume the contrary. Then considering the $\sigma = (1)$ term and replacing x^*_i with x_{2n+i}, we see that

$$\alpha x_1 \beta x_{2n+2} \alpha x_3 \beta x_{2n+4} \cdots \alpha x_{2n-1} \beta x_{4n}$$

is a GPI for FG. Thus, letting I_1 be the ideal generated by α and I_2 the ideal generated by β, we find that $(I_1 I_2)^n = 0$. Since FG is prime, $I_1 = 0$ or $I_2 = 0$. We are done. $\square$

Lemma 4.4.5. *Suppose G is torsion, FG is prime, $\alpha \in (FG)^+$ and n is a positive integer. If $(\alpha\beta)^n = 0$ for all $\beta \in (FG)^+$, then $\alpha = 0$.*

Proof. For any $\gamma \in FG$, we have $\gamma\alpha\gamma^* \in (FG)^+$, hence $(\alpha\gamma\alpha\gamma^*)^n = 0$. Apply the previous lemma with $\alpha = \beta$. $\square$

We need a slightly different form for our group identity than that provided by Lemma 2.1.3.

Lemma 4.4.6. *Let R be a ring with involution. If $\mathcal{U}^+(R)$ satisfies a group identity, then it satisfies a group identity of the form $x^{i_1} y^{j_1} \cdots x^{i_k} y^{j_k} = 1$, where each exponent is nonzero, $i_1 > 0$ and $j_k < 0$.*

Proof. We can always make $i_1 > 0$ and $j_k < 0$ by replacing x with x^{-1} or y with y^{-1}, if necessary. Thus, we need not concern ourselves with that point. In view of Lemma 2.1.3, our identity can be assumed to be of the form $x^{i_1} y^{j_1} \cdots y^{j_{k-1}} x^{i_k}$, with no exponent equal to zero. If $k = 1$, then we replace x with xyx and we see that $\mathcal{U}^+(R)$ satisfies $(x^2 y)^{i_1} = 1$. Therefore, we can assume that $k \geq 2$. Now, $\mathcal{U}^+(R)$ satisfies $x^{i_1 + i_k} y^{j_1} \cdots x^{i_{k-1}} y^{j_{k-1}}$. If $i_1 + i_k \neq 0$, then we are done. Otherwise, we exchange x and y and repeat. Eventually, we must obtain the desired form or else an identity of the form $x^r = 1$, with $r \neq 0$. But we dealt with the latter case above. We are done. $\square$

Lemma 4.4.7. *Suppose char $F \neq 2$, G is torsion, FG is prime and $\mathcal{U}^+(FG)$ satisfies a group identity. Then there exists a positive integer k such that for all square-zero $\alpha \in FG$, we have $(\alpha^*\alpha)^k = 0$.*

Proof. Let the group identity be $w(x, y) = x^{i_1} \cdots y^{j_k}$ as in the preceding lemma. Now, $(1 + \alpha)(1 + \alpha^*)$ and $(1 + \alpha^*)(1 + \alpha)$ lie in $\mathcal{U}^+(FG)$ (with inverses $(1 - \alpha^*)(1 - \alpha)$ and $(1 - \alpha)(1 - \alpha^*)$, respectively). Thus,

$$((1 + \alpha)(1 + \alpha^*))^{i_1}((1 + \alpha^*)(1 + \alpha))^{j_1} \cdots ((1 + \alpha^*)(1 + \alpha))^{j_k} = 1.$$

Here, we have a product of terms of the form $1 \pm \alpha$ and $1 \pm \alpha^*$, with no more than two identical terms adjacent to each other. Also, $1 + \alpha$ is never adjacent to $1 - \alpha$, nor is $1 + \alpha^*$ adjacent to $1 - \alpha^*$. As $(1 + \alpha)^2 = 1 + 2\alpha$ and $(1 + \alpha^*)^2 = 1 + 2\alpha^*$, we get

$$(1 + \alpha)(1 + \lambda_1 \alpha^*)(1 + \lambda_2 \alpha)(1 + \lambda_3 \alpha^*) \cdots (1 + \lambda_m \alpha)(1 - \alpha^*) = 1,$$

where each $\lambda_i \in \{\pm 1, \pm 2\}$ (and the first and last terms follow from the fact that $i_1 > 0$ and $j_k < 0$).

Expanding this expression and discarding all terms in which α^2 or $(\alpha^*)^2$ appears, we get a sum of terms of the form $\pm 2^r (\alpha^*)^n (\alpha \alpha^*)^u \alpha^l$, where $r, u \geq 0$, $l, n \in \{0, 1\}$. Multiplying on the left by α^* and on the right by α, we eliminate all of the terms in which $l = 1$ or $n = 1$. That is, we have a polynomial $f(t)$ in $\alpha^* \alpha$. The leading term is evidently $-\lambda_1 \lambda_2 \cdots \lambda_m (\alpha^* \alpha)^d$ for some d. Since each $\lambda_i \in \{\pm 1, \pm 2\}$, the coefficient is not zero. Write $f(t) = \sum_{i=1}^{d} \mu_i t^i$. (The constant term is clearly zero.) Note that this polynomial does not depend upon the particular choice of α. Take any $\beta \in FG$. Then $(\alpha \beta \alpha)^2 = 0$, hence

$$0 = f((\alpha \beta \alpha)^* \alpha \beta \alpha) = \sum_{i=1}^{d} \mu_i (\alpha^* \beta^* \alpha^* \alpha \beta \alpha)^i.$$

Thus, FG satisfies the $*$-GPI

$$\sum_{i=1}^{d} \mu_i (\alpha^* x_1^* \alpha^* \alpha x_1 \alpha)^i.$$

Linearizing this expression, all terms except those of the highest degree will vanish, since eventually, we must perform a linearization step with a variable with respect to which the term is already linear. Dividing by μ_d, the linearization is

$$\sum_{\sigma \in S_{2d}} \alpha^* x_{\sigma(1)}^* \alpha^* \alpha x_{\sigma(2)} \alpha \alpha^* x_{\sigma(3)}^* \alpha^* \alpha x_{\sigma(4)} \alpha \cdots \alpha^* x_{\sigma(2d-1)}^* \alpha^* \alpha x_{\sigma(2d)} \alpha.$$

If this $*$-GPI is nondegenerate, then by Lemma 4.4.3, $G = 1$ and there is nothing to do. Otherwise, looking at the $\sigma = (1)$ term and replacing x_i^* with x_{2d+i} for all i, we see that FG must satisfy

$$\alpha^* x_{2d+1} \alpha^* \alpha x_2 \alpha \alpha^* x_{2d+3} \alpha^* \alpha x_4 \alpha \cdots \alpha^* x_{4d-1} \alpha^* \alpha x_{2d} \alpha = 0.$$

Substituting $x_i = \alpha^*$ for $i \leq 2d$ and $x_i = \alpha$ for $i > 2d$, we get $(\alpha^* \alpha)^{3d} = 0$. We are done. $\qquad \square$

Lemma 4.4.8. *Suppose char $F \neq 2$, G is torsion, FG is prime and $\mathscr{U}^+(FG)$ satisfies a group identity. Then, for all square-zero $\alpha, \beta \in (FG)^+$, we have $\alpha \beta \alpha = 0$.*

Proof. Let the group identity be $w(x, y) = x^{i_1} \cdots y^{j_k}$ as in Lemma 4.4.6. Notice that $(1 + \beta)(1 + \alpha)(1 + \beta), (1 + \alpha)(1 + \beta)(1 + \alpha) \in \mathscr{U}^+(FG)$. Thus,

$$1 = w((1 + \beta)(1 + \alpha)(1 + \beta), (1 + \alpha)(1 + \beta)(1 + \alpha))$$
$$= (1 + \beta)(1 + \alpha)(1 + \lambda_1 \beta)(1 + \lambda_2 \alpha) \cdots (1 + \lambda_m \alpha)(1 - \beta)(1 - \alpha)$$

for some $\lambda_i \in \{\pm 1, \pm 2\}$. (This occurs because all of the terms in our product are of the form $1 \pm \alpha$ or $1 \pm \beta$ with no more than two consecutive identical terms and $1 + \alpha$ not appearing next to $1 - \alpha$, and $1 + \beta$ not appearing next to $1 - \beta$.)

Multiplying on the left by α and on the right by β we obtain, as in the proof of the previous lemma, a polynomial $f(t) = \sum_{i=1}^d \mu_i t^i$ with $\mu_d \neq 0$ and $f(\alpha\beta) = 0$ for all square-zero $\alpha, \beta \in (FG)^+$. Take any $\delta \in FG$. Then $\alpha(\delta + \delta^*)\alpha$ and $\beta(\delta + \delta^*)\beta$ are symmetric and square-zero. Thus,

$$0 = f(\alpha(\delta + \delta^*)\alpha\beta(\delta + \delta^*)\beta)$$
$$= \sum_{i=1}^d \mu_i(\alpha(\delta + \delta^*)\alpha\beta(\delta + \delta^*)\beta)^i.$$

That is, FG satisfies the $*$-GPI

$$\sum_{i=1}^d \mu_i(\alpha(x_1 + x_1^*)\alpha\beta(x_1 + x_1^*)\beta)^i.$$

As we observed in the previous lemma, when we linearize, only the terms of highest degree remain. Dividing out the leading coefficient, we find that FG satisfies the multilinear $*$-GPI

$$\sum_{\sigma \in S_{2d}} \sum_{\varepsilon_i \in \{1,*\}} \alpha x_{\sigma(1)}^{\varepsilon_1} \alpha \beta x_{\sigma(2)}^{\varepsilon_2} \beta \alpha x_{\sigma(3)}^{\varepsilon_3} \alpha \beta x_{\sigma(4)}^{\varepsilon_4} \beta \cdots \alpha x_{\sigma(2d-1)}^{\varepsilon_{2d-1}} \alpha \beta x_{\sigma(2d)}^{\varepsilon_{2d}} \beta.$$

If this last identity is nondegenerate, then $G = 1$ and there is nothing to do. Otherwise, replacing x_i^* with x_{2d+i} for all i, and then looking at the monomial in which the variables $x_1, x_2, \ldots, x_{2d}$ occur, in that order, we see that FG satisfies

$$\alpha x_1 \alpha \beta x_2 \beta \cdots \alpha x_{2d-1} \alpha \beta x_{2d} \beta = 0.$$

Letting $x_i = \beta$ when i is odd and $x_i = \alpha$ when i is even, we get $(\alpha\beta)^{3d} = 0$. Now, if $\gamma \in (FG)^+$, then $\alpha\gamma\alpha$ is symmetric and square-zero. Thus, $(\alpha\gamma\alpha\beta)^{3d} = 0$, hence $(\alpha\beta\alpha\gamma)^{3d+1} = 0$. Now apply Lemma 4.4.5.   □

But we need to extend this slightly.

Lemma 4.4.9. *Suppose char $F \neq 2$, G is torsion, FG is prime and $\mathcal{U}^+(FG)$ satisfies a group identity. If $\alpha, \beta \in (FG)^+$ are such that α is square-zero and β is nilpotent, then $\alpha\beta\alpha = 0$.*

Proof. Let m be the smallest positive integer such that $\beta^m = 0$. We proceed by induction on m. If $m = 1$, there is nothing to do. If $m = 2$, then the preceding lemma does the job. So, let $m > 2$. Then $(\beta^2)^{m-1} = 0$. Thus, by our inductive hypothesis, $\alpha\beta^2\alpha = 0$. Take any $\gamma \in (FG)^+$. Then $\beta\alpha\gamma\alpha\beta$ is symmetric and square-zero. Therefore, by the $m = 2$ case, $\alpha\beta\alpha\gamma\alpha\beta\alpha = 0$. In particular, $(\alpha\beta\alpha\gamma)^2 = 0$. Thus, by Lemma 4.4.5, $\alpha\beta\alpha = 0$.   □

This allows us to give the

Proof of Proposition 4.4.1. The characteristic zero case follows from Propositions 2.4.2 and 4.3.1, so assume that char $F = p > 2$. Let us first show that G is a p-group. Suppose that $1 \neq g \in G$ is a p'-element. Then $e = \frac{1}{o(g)}\hat{g}$ is a symmetric idempotent. Hence, for all $\alpha \in FG$ we have $(e\alpha(1 - e))^2 = 0$. Thus, by Lemma 4.4.7, there exists an n so that

$$((e\alpha(1 - e))^* e\alpha(1 - e))^n = 0$$

for all $\alpha \in FG$. That is,

$$((1 - e)\alpha^* e\alpha(1 - e))^n = 0.$$

Thus, $(e\alpha(1 - e)\alpha^*)^{n+1} = 0$ for any $\alpha \in FG$. By Lemma 4.4.4, $e = 0$ or 1, both of which are impossible. Thus, G is a p-group.

Let g be any element of order p in G. In view of the preceding lemma, the proof of Lemma 2.4.1 shows us that $\langle g \rangle$ is a normal subgroup. But FG is prime, and we have a contradiction.   □

Combining Proposition 4.4.1 with Propositions 4.3.2 and 4.3.3, we obtain

Corollary 4.4.10. *Let G be a locally finite group not containing Q_8 and char $F = p > 2$. If $\mathscr{U}^+(FG)$ is nilpotent, then G is nilpotent and p-abelian.*

Corollary 4.4.11. *Let G be a locally finite group containing Q_8 and char $F = p > 2$. If $\mathscr{U}^+(FG)$ is nilpotent, then $G \simeq Q_8 \times E \times P$, where E is an elementary abelian 2-group and P is a finite p-group.*

4.5 Group Rings of Torsion Groups

We must now show that if $\mathscr{U}^+(FG)$ is nilpotent (for torsion G), then G is locally finite.

Let $\langle x_1, x_2, \ldots \rangle$ be the free group on generators $x_1, x_2, \ldots$. Then recall that a semigroup identity is a group identity of the form

$$x_{i_1} x_{i_2} \cdots x_{i_m} = x_{j_1} x_{j_2} \cdots x_{j_n},$$

where the two words are not identical in $\langle x_1, x_2, \ldots \rangle$. The following lemma is from Mal'cev [77].

Lemma 4.5.1. *Every nilpotent group satisfies a semigroup identity.*

Proof. Let us define words u_i and v_i in $\langle x_1, x_2, \ldots \rangle$ as follows. We let $u_0 = x_1$ and $v_0 = x_2$. Then, for each $i \geq 0$, let $u_{i+1} = u_i x_{i+3} v_i$ and $v_{i+1} = v_i x_{i+3} u_i$. Evidently u_i and v_i are different words for all i. Let G be nilpotent of class n. We claim that G satisfies $u_n = v_n$. Our proof is by induction on n. If $n = 1$, then $u_1 = x_1 x_3 x_2$ and

$v_1 = x_2 x_3 x_1$. Clearly an abelian group satisfies $u_1 = v_1$. Assume that $n > 1$ and the claim holds for smaller n. Since $\bar{G} = G/\zeta(G)$ has smaller nilpotency class, it satisfies $u_{n-1} = v_{n-1}$. Choose any $g_i \in G$. Then $u_{n-1}(\bar{g}_1, \ldots, \bar{g}_{n+1}) = v_{n-1}(\bar{g}_1, \ldots, \bar{g}_{n+1})$, hence $u_{n-1}(g_1, \ldots, g_{n+1}) = v_{n-1}(g_1, \ldots, g_{n+1})z$ for some $z \in \zeta(G)$. Thus,

$$
\begin{aligned}
u_n(g_1, \ldots, g_{n+2}) &= u_{n-1}(g_1, \ldots, g_{n+1})g_{n+2}v_{n-1}(g_1, \ldots, g_{n+1}) \\
&= v_{n-1}(g_1, \ldots, g_{n+1})zg_{n+2}v_{n-1}(g_1, \ldots, g_{n+1}) \\
&= v_{n-1}(g_1, \ldots, g_{n+1})g_{n+2}v_{n-1}(g_1, \ldots, g_{n+1})z \\
&= v_{n-1}(g_1, \ldots, g_{n+1})g_{n+2}u_{n-1}(g_1, \ldots, g_{n+1}) \\
&= v_n(g_1, \ldots, g_{n+2}).
\end{aligned}
$$

We are done. $\square$

Lemma 4.5.2. *Let F be a field and R an F-algebra with involution such that $\mathscr{U}^+(R)$ is nilpotent. If I is a $*$-invariant nil ideal of R, then I satisfies a polynomial identity.*

Proof. By Lemmas 4.1.2 and 4.5.1, $\langle \mathscr{U}^+(R) \rangle$ satisfies a semigroup identity, say $x_{i_1} \cdots x_{i_m} = x_{j_1} \cdots x_{j_n}$. Choose any $r_i \in I^+$. Then each $1 + r_i \in \mathscr{U}^+(R)$, hence

$$(1 + r_{i_1}) \cdots (1 + r_{i_m}) = (1 + r_{j_1}) \cdots (1 + r_{j_n}).$$

Expanding, we obtain a polynomial identity for I^+. (Since $x_{i_1} \cdots x_{i_m} \neq x_{j_1} \cdots x_{j_n}$, the leading terms on each side are not equal, so this is not the zero polynomial.) Thus, I satisfies a $*$-polynomial identity. By Proposition 2.1.2, I satisfies a polynomial identity. $\square$

Let us now eliminate the centreless groups.

Lemma 4.5.3. *Let char $F = p > 2$ and let G be a torsion group with $\zeta(G) = 1$. If $\mathscr{U}^+(FG)$ is nilpotent, then G is the trivial group.*

Proof. First, suppose that FG is semiprime. If FG is prime, then Proposition 4.4.1 does the job, so let N be a nontrivial finite normal subgroup of G. By Proposition 1.2.9, N is a p'-group. Take $h \in N$ of prime order, and take any $g \in G$. Then $\langle N, g \rangle$ is finite. By Corollaries 4.4.10 and 4.4.11, either $\langle N, g \rangle$ is nilpotent and p-abelian or $\langle N, g \rangle \simeq Q_8 \times E \times P$, where E is an elementary abelian 2-group and P is a finite p-group. Either way, an element of prime order different from p is central in $\langle N, g \rangle$. Thus, h commutes with g. But g was arbitrary, hence $h \in \zeta(G)$, giving us a contradiction.

Now, suppose that $N(FG)$ is nilpotent. By Proposition 1.2.22, $\phi_p(G)$ is finite, and by Lemma 4.3.9 it is a p-group. Let $\bar{G} = G/\phi_p(G)$. By Lemma 2.3.6, $\mathscr{U}^+(F\bar{G})$ is nilpotent. Also, by Proposition 1.2.9, $F\bar{G}$ is semiprime. We claim that $\zeta(\bar{G}) = 1$. Choose $\bar{z} \in \zeta(\bar{G})$. Then $\bar{z} \in \phi(\bar{G})$, hence $z \in \phi(G)$. Write $z = z_1 z_2$, where z_1 is a p'-element and $z_2 \in \phi_p(G)$. Then $\bar{z} = \bar{z}_1$, so we may assume that z is a p'-element. If $g \in G$, then z and z^g are p'-elements in $\phi(G)$. By Lemma 1.4.4, $\phi(G)$ is locally finite. Thus, by Corollaries 4.4.10 and 4.4.11, $\phi(G)$ is nilpotent. It follows that $(z, g) =$

$z^{-1}z^g$ is a p'-element. But since $\bar{z} \in \zeta(\bar{G})$, $(z,g) \in \phi_p(G)$, which is a p-group. Thus, $z \in \zeta(G) = 1$, as claimed. By the semiprime case, $\bar{G} = 1$, hence $G = \phi_p(G)$ is finite. But a centreless finite p-group is trivial.

Finally, suppose that $N(FG)$ is not nilpotent. By the last lemma, $N(FG)$ satisfies a polynomial identity. But then Lemma 1.2.16 tells us that FG satisfies a nondegenerate multilinear GPI. By Proposition 1.2.15, $(G : \phi(G)) < \infty$ and $|(\phi(G))'| < \infty$. In particular, G is locally finite. By Corollaries 4.4.10 and 4.4.11, G is nilpotent. But then G is only centreless if it is trivial. □

Lemma 4.5.4. *Let G be a group and char $F \neq 2$. Choose any ordinal α. Let $N_0 \leq N_1 \leq \cdots \leq N_\alpha$ be an ascending chain of normal subgroups of G such that $\mathscr{U}^+(F(G/N_\beta))$ satisfies the group identity $w(x_1, \ldots, x_n) = 1$ for all $\beta < \alpha$. If $N_\alpha = \bigcup_{\beta < \alpha} N_\beta$, then $\mathscr{U}^+(F(G/N_\alpha))$ satisfies $w(x_1, \ldots, x_n) = 1$.*

Proof. Suppose this is not the case. Let $\bar{G} = G/N_\alpha$. Choose $\bar{\gamma}_i \in \mathscr{U}^+(F\bar{G})$ such that $w(\bar{\gamma}_1, \ldots, \bar{\gamma}_n) \neq 1$. As $\varepsilon_{N_\alpha}((FG)^+) = (F\bar{G})^+$, choose $\gamma_i \in (FG)^+$ such that $\varepsilon_{N_\alpha}(\gamma_i) = \bar{\gamma}_i$.

If we can show that there exists a $\beta < \alpha$ such that $\varepsilon_{N_\beta}(\gamma_i)$ is a unit for all i, then we will have $w(\varepsilon_{N_\beta}(\gamma_1), \ldots, \varepsilon_{N_\beta}(\gamma_n)) = 1$, hence $w(\bar{\gamma}_1, \ldots, \bar{\gamma}_n) = 1$, and we will have a contradiction.

Let $\bar{\delta}_i = \bar{\gamma}_i^{-1}$. We choose $\delta_i \in (FG)^+$ such that $\varepsilon_{N_\alpha}(\delta_i) = \bar{\delta}_i$. Then $\gamma_i \delta_i - 1$ and $\delta_i \gamma_i - 1$ lie in $\Delta(G, N_\alpha)$. We can therefore find a finite subset S of N_α such that all $\gamma_i \delta_i - 1$ and $\delta_i \gamma_i - 1$ can be written as a linear combination of terms of the form $g(s-1)$ with $g \in G$, $s \in S$. As $N_\alpha = \bigcup_{\beta < \alpha} N_\beta$, each s lies in some N_{β_s}, with $\beta_s < \alpha$. Let β be the largest of the β_s. Then $\gamma_i \delta_i - 1, \delta_i \gamma_i - 1 \in \Delta(G, N_\beta)$. That is, $\varepsilon_{N_\beta}(\gamma_i)$ is a unit. We are done. □

Recall that the (transfinitely extended) upper central series of a group G is defined as follows. We let $\zeta_0(G) = 1$ and for each ordinal α let $\zeta_{\alpha+1}(G)/\zeta_\alpha(G) = \zeta(G/\zeta_\alpha(G))$. If α is a limit ordinal, then $\zeta_\alpha(G) = \bigcup_{\beta < \alpha} \zeta_\beta(G)$. Eventually, we must find an ordinal α so that $\zeta_\alpha(G) = \zeta_{\alpha+1}(G)$. In this case, we call $\zeta_\alpha(G)$ the hypercentre of G. If G is equal to its own hypercentre, then G is said to be hypercentral.

As promised, we now have

Lemma 4.5.5. *Let G be a torsion group and char $F \neq 2$. If $\mathscr{U}^+(FG)$ is nilpotent, then G is locally finite.*

Proof. As the characteristic zero case follows from Proposition 2.4.2, assume that char $F = p > 2$. If there is a group G for which the result fails, then there is one that is finitely generated. Thus, assume that G is countable. Let $\mathscr{U}^+(FG)$ satisfy $(x_1, \ldots, x_n) = 1$. We claim that for each ordinal α, $\mathscr{U}^+(F(G/\zeta_\alpha(G)))$ also satisfies $(x_1, \ldots, x_n) = 1$. Our proof is by transfinite induction. If α is a limit ordinal, then Lemma 4.5.4 does the job. Thus, it suffices to assume that this holds for α but not $\alpha + 1$. In fact, then, it suffices to assume that $\mathscr{U}^+(FG)$ satisfies $(x_1, \ldots, x_n) = 1$ but $\mathscr{U}^+(F(G/\zeta(G)))$ does not.

If $\zeta(G)$ is finite, then let H be its set of p-elements. We see from Lemma 2.3.6 that $\mathscr{U}^+(F(G/H))$ satisfies $(x_1,\ldots,x_n) = 1$ and then, by Lemma 2.4.6, so does

$$\mathscr{U}^+(F((G/H)/(\zeta(G)/H))) = \mathscr{U}^+(F(G/\zeta(G))).$$

Thus, assume that $\zeta(G)$ is infinite. Let $\zeta(G) = \{z_i : i \geq 1\}$. Then define an ascending chain of central subgroups of G in the following way: namely, $N_0 = 1$ and $N_i = \langle N_{i-1}, z_i \rangle$. Evidently, each N_i is finite, hence, as we saw above, each $\mathscr{U}^+(F(G/N_i))$ satisfies $(x_1,\ldots,x_n) = 1$. Since $\zeta(G) = \bigcup_{i=1}^{\infty} N_i$, the preceding lemma tells us that $\mathscr{U}^+(F(G/\zeta(G)))$ satisfies $(x_1,\ldots,x_n) = 1$, and our claim is proved.

In particular, let $\zeta_\alpha(G)$ be the hypercentre. Now $G/\zeta_\alpha(G)$ is centreless; thus, by Lemma 4.5.3, G is hypercentral. But every hypercentral group is locally nilpotent (see, for instance, [89, 12.2.4]). As a torsion nilpotent group is locally finite, we are done.   □

We now present the main results of the author in [67].

Theorem 4.5.6. *Let F be a field of characteristic $p \neq 2$ and G a torsion group not containing Q_8. Then the following are equivalent:*

(i) $\mathscr{U}^+(FG)$ is nilpotent;
(ii) $\mathscr{U}(FG)$ is nilpotent;
(iii) G is nilpotent and p-abelian.

Proof. If $p = 0$, use Proposition 2.4.2. Otherwise, combine Lemma 4.5.5, Corollary 4.4.10 and Proposition 4.2.6.   □

Theorem 4.5.7. *Let F be a field of characteristic $p \neq 2$ and G a torsion group containing Q_8. Then $\mathscr{U}^+(FG)$ is nilpotent if and only if either*

1. $p = 0$ and $G \simeq Q_8 \times E$, where E is an elementary abelian 2-group, or
2. $p > 2$ and $G \simeq Q_8 \times E \times P$, where E is an elementary abelian 2-group and P is a finite p-group.

Proof. If $p = 0$, then Proposition 2.4.2 and Lemma 2.1.1 do all of the work. Let us assume that $p > 2$. In view of Lemma 4.5.5 and Corollary 4.4.11, we only need to check the sufficiency. We claim that, for any $n \geq 2$ and all $\alpha_1,\ldots,\alpha_n \in \mathscr{U}^+(FG)$, we have $(\alpha_1,\ldots,\alpha_n) - 1 \in ((FG)^+)^{(n)}$. Our proof is by induction on n. If $n = 2$, then

$$(\alpha_1,\alpha_2) - 1 = \alpha_1^{-1}\alpha_2^{-1}[\alpha_1,\alpha_2] \in ((FG)^+)^{(2)}.$$

Supposing that the claim holds for n, we have

$$\begin{aligned}
(\alpha_1,\ldots,\alpha_n,\alpha_{n+1}) - 1 &= ((\alpha_1,\ldots,\alpha_n),\alpha_{n+1}) - 1 \\
&= (\alpha_1,\ldots,\alpha_n)^{-1}\alpha_{n+1}^{-1}[(\alpha_1,\ldots,\alpha_n),\alpha_{n+1}] \\
&= (\alpha_1,\ldots,\alpha_n)^{-1}\alpha_{n+1}^{-1}[(\alpha_1,\ldots,\alpha_n) - 1,\alpha_{n+1}],
\end{aligned}$$

which lies in $((FG)^+)^{(n+1)}$, by our inductive hypothesis. By Lemma 3.3.5, $(FG)^+$ is strongly Lie nilpotent. Thus, choosing an n such that $((FG)^+)^{(n)} = 0$, we have $(\alpha_1,\ldots,\alpha_n) = 1$ for all $\alpha_i \in \mathscr{U}^+(FG)$. The proof is complete.   □

Comparing the last two theorems with Theorems 3.3.1 and 3.3.6 gives us the following result.

Corollary 4.5.8. *Let F be a field of characteristic different from 2 and G a torsion group. Then $\mathscr{U}^+(FG)$ is nilpotent if and only if $(FG)^+$ is Lie nilpotent.*

4.6 Semiprime Group Rings of Nontorsion Groups

Let us now discuss the results of Lee et al. [69] concerning groups G containing elements of infinite order such that $\mathscr{U}^+(FG)$ is nilpotent. These results depend upon the classification, given in Chapter 2, of groups G such that $\mathscr{U}^+(FG)$ satisfies a group identity. As the assumption there was that F is an infinite field, we impose that restriction here as well. We consider the semiprime case in this section and the general case in the next section.

Let F be an infinite field of characteristic $p \neq 2$ and let G be a group containing an element of infinite order. Suppose that $\mathscr{U}^+(FG)$ is nilpotent and FG is semiprime. In view of Theorem 2.5.6, we know that the torsion elements of G form a subgroup, T. Furthermore, if $p = 0$, then T is abelian or a Hamiltonian 2-group; if $p > 2$, then T is an abelian p'-group. In addition, every idempotent of FT is central in FG. One more restriction is required.

Let H be a finite subgroup of T and suppose that FH has a nonsymmetric primitive idempotent e. Of course, e is central in FG. If T is a Hamiltonian 2-group, then by Lemma 2.1.1, the centre of FT is equal to its set of symmetric elements. Thus, we have a contradiction, and we may assume that T is abelian.

Let N be the kernel of the homomorphism $G \to \mathscr{U}(FGe)$ given by $g \mapsto ge$. Since the support of e is contained in H, the support of ge is contained in gH. Thus, $N \leq H$. We claim that $\bar{G} = G/N$ is nilpotent. Indeed, the Wedderburn decomposition of FH is

$$FH = FHe \oplus FHe^* \oplus \cdots$$

and, since e is central, we can write

$$FG = FGe \oplus FGe^* \oplus \cdots.$$

If $g \in G$, then

$$ge + g^{-1}e^* + (1 - (e + e^*)) \in \mathscr{U}^+(FG).$$

Thus, if $\mathscr{U}^+(FG)$ satisfies $(x_1, \ldots, x_n) = 1$, then taking any $g_1, \ldots, g_n \in G$, and looking only at the FGe component, we get

$$(g_1 e, \ldots, g_n e) = e,$$

hence

$$(g_1, \ldots, g_n)e = e.$$

That is, $(g_1,\ldots,g_n) \in N$ and, therefore, $(\bar{g}_1,\ldots,\bar{g}_n) = 1$ for all $\bar{g}_i \in \bar{G}$, as claimed. In particular, G/T is nilpotent.

We also claim that $\bar{H} \le \zeta(\bar{G})$. To prove this, we first note that $\bar{e}$ is not symmetric. Indeed, if $\bar{e} = \bar{e}^*$, then $e - e^* \in \Delta(G,N)$. Thus, by Lemma 1.3.8, $\hat{N}(e - e^*) = 0$. But by definition of N, $\hat{N}e = |N|e$. Also, since $\hat{N}$ is symmetric, $\hat{N}e^* = (\hat{N}e)^* = |N|e^*$. That is, $|N|(e - e^*) = 0$. Since $|N|$ is a unit, $e - e^* = 0$, giving us a contradiction. Also, $\bar{e}$ is a primitive idempotent of $F\bar{H}$. Since, by Lemma 2.4.6, $\mathscr{U}^+(F\bar{G})$ is nilpotent, we can replace G with $\bar{G}$ and therefore assume that $g \mapsto ge$ is injective. Since FH is commutative, FHe is a field. Thus H, being isomorphic to He (a finite subgroup of a field), is cyclic.

For any $0 \ne \alpha e \in FHe$, we have $\alpha e + \alpha^* e^* + (1 - (e + e^*)) \in \mathscr{U}^+(FG)$. We also have $ge + g^{-1}e^* + (1 - (e + e^*)) \in \mathscr{U}^+(FG)$. Thus, working in the first component, and using the notation of Lemma 4.2.8, we have

$$(\alpha e, \underbrace{g,\ldots,g}_{n-1 \text{ times}}) = e.$$

Since $\frac{1}{|H|}\hat{H}$ is an idempotent, it is central; hence, H is normal in G. Furthermore, since H is finite, there exists an $m > 0$ such that g^m centralizes H and, therefore, FHe. By Lemma 4.2.8, g centralizes FHe.

Writing $H = \langle h \rangle$, we know that the image of h in each Wedderburn component of FH is an mth root of unity, where $o(h) = m$. In particular, $H \to He$ is injective; hence, he is a primitive mth root of unity, ξ. Regarding every Wedderburn component as a subfield of $F(\xi)$, we have

$$h = (\xi, \xi^{i_2}, \ldots, \xi^{i_l})$$

for various integers i_j. Now, $h^g = h^i$ for some $i \ge 0$, hence

$$h^g = (\xi^i, \xi^{ii_2}, \ldots, \xi^{ii_l}).$$

But g centralizes FHe. Thus, $\xi = \xi^i$ and therefore $h = h^g$. That is, h is central in G, as claimed.

The first of the main results from [69] is

Theorem 4.6.1. *Let F be an infinite field of characteristic $p \ne 2$ and G a group containing an element of infinite order, such that FG is semiprime. If $\mathscr{U}^+(FG)$ is nilpotent, then the torsion elements of G form a (normal) subgroup T such that*

1. *if $p = 0$, then T is abelian or a Hamiltonian 2-group;*
2. *if $p > 2$, then T is an abelian p'-group;*
3. *every idempotent of FT is central in FG; and*
4. *if H is a finite subgroup of G such that FH has a nonsymmetric primitive idempotent e, then G/T is nilpotent and H/N is central in G/N, where N is the kernel of $g \mapsto ge$.*

Conversely, if G/T is a u.p. group and FG satisfies the four conditions above, then $\mathscr{U}^+(FG)$ is nilpotent.

Proof. We proved the necessity above, so let us check the sufficiency of the conditions. If every idempotent of FT is symmetric, then by Remark 2.5.7, $\mathscr{U}^+(FG)$ is commutative. Therefore, we may assume that FT has a nonsymmetric idempotent. In particular, for some finite subgroup H of T, FH has a nonsymmetric primitive idempotent. Thus, G/T is nilpotent. Let G/T satisfy $(x_1,\ldots,x_n)=1$. We claim that $\mathscr{U}^+(FG)$ satisfies $(x_1,\ldots,x_{n+1})=1$. Suppose this is not the case, and take symmetric units $\alpha_1,\ldots,\alpha_{n+1}$ such that $(\alpha_1,\ldots,\alpha_{n+1})\neq 1$. Let X be a transversal of T in G containing 1. For each i write $\alpha_i=\sum_j \beta_{ij}g_j$, with $\beta_{ij}\in FT$ and g_j in X. Similarly, write $\alpha_i^{-1}=\sum_j \beta'_{ij}g_j$. For all g_i and g_j appearing in the above expressions, let $g_ig_j=t_{ij}h_{ij}$, with $t_{ij}\in T$, $h_{ij}\in X$. Finally, for all elements g_i of X appearing in the above expressions, let $(g_1,\ldots,g_n)=hg$, with $h\in T$, $g\in X$. Then let E be the finitely generated (hence finite) subgroup of T generated by the supports of the β_{ij} and β'_{ij}, the t_{ij} and the h.

It suffices to show that for each primitive idempotent e of FE, we have

$$(\alpha_1,\ldots,\alpha_{n+1})e=e,$$

since then summing over all of the Wedderburn components gives us a contradiction. By Remark 1.4.10, each $\alpha_ie=\lambda_ig_i$, where $\lambda_i\in\mathscr{U}(FEe)$ and g_i is among the elements of X described above.

Suppose $e^*=e$. Then $\alpha_ie\in(FG)^+$, hence $(\lambda_ig_i)^*=\lambda_ig_i$. That is, $g_i^{-1}\lambda_i^*=\lambda_ig_i$, hence $(\lambda_i^*)^{g_i}=\lambda_ig_i^2$. Now surely $(\lambda_i^*)^{g_i}\in FE$, but $\lambda_ig_i^2\notin FE$ unless g_i^2 is a torsion element. But this can only happen if $g_i=1$. Thus, $\alpha_ie\in(FE)^+$. But FE is abelian or a Hamiltonian 2-group. Therefore, by Lemma 2.1.1, $(FE)^+$ is commutative. Thus, the α_ie commute.

On the other hand, suppose that $e^*\neq e$. Our conditions then guarantee that G/T satisfies $(x_1,\ldots,x_n)=1$ and E/N is central in G/N, where N is the kernel of $g\mapsto ge$. Then FEe is central in FGe. Thus,

$$(\alpha_1e,\ldots,\alpha_ne)=(\lambda_1g_1,\ldots,\lambda_ng_n)=(g_1,\ldots,g_n)e.$$

By definition of E, we have $(g_1,\ldots,g_n)\in E$, hence $(\alpha_1e,\ldots,\alpha_ne)\in FEe$. But again, FEe is central, hence $(\alpha_1e,\ldots,\alpha_ne,\alpha_{n+1}e)=e$. We are done. $\qquad\square$

In particular, the characteristic zero case is now completely resolved. Groups containing the quaternions will only appear in this case, due to

Corollary 4.6.2. *Let G be a group containing an element of infinite order, and let F be an infinite field of characteristic $p\neq 2$. If G contains the quaternions, and $\mathscr{U}^+(FG)$ is nilpotent, then $p=0$.*

Proof. If FG is semiprime, there is nothing more to say. Otherwise, by Theorems 2.6.4, 2.6.5 and 2.6.11, we know that the torsion elements form a subgroup, T. By Theorem 4.5.7, $T\simeq Q_8\times E\times P$, where E is an elementary abelian 2-group and P is a finite p-group. By Lemma 2.3.6, $\mathscr{U}^+(F(G/P))$ is nilpotent. But by Proposition 1.2.9, $F(G/P)$ is semiprime. Thus, by the preceding theorem, since G/P contains the quaternions, $p=0$, as required. $\qquad\square$

4.7 The General Case for Nontorsion Groups

Let us now suppose that $\mathcal{U}^+(FG)$ is nilpotent, where F is an infinite field of characteristic $p > 2$ and G is a nontorsion group containing p-elements. In view of Theorems 2.6.4, 2.6.5 and 2.6.11, the torsion elements form a normal subgroup T of G. Thus, by Theorem 4.5.6 and Corollary 4.6.2, T is nilpotent and T' is a finite p-group. In particular, $T = A \times P$, where P is a p-group and A is an abelian p'-group.

Let us first suppose that P is finite. Then by Lemma 2.3.6, $\mathcal{U}^+(F(G/P))$ is nilpotent. But by Proposition 1.2.9, $F(G/P)$ is semiprime. Thus, Theorem 4.6.1 applies to $F(G/P)$. Two additional conditions are required.

Lemma 4.7.1. *Let F be an infinite field of characteristic $p > 2$ and G a nontorsion group with $1 < |P| < \infty$ such that $\mathcal{U}^+(FG)$ is nilpotent. If $h \in G$ has finite p'-order and g is any element of G, then $h^g = h$ or h^{-1}.*

Proof. Let $H = \langle h \rangle$. Then FH is semisimple. We will prove that for each primitive idempotent e of FH we have

$$g^{-1}(h + h^{-1})ge = (h + h^{-1})e$$

for all $g \in G$. This will complete the proof, since summing over all e, we get

$$g^{-1}hg + g^{-1}h^{-1}g = h + h^{-1},$$

hence $h^g = h$ or h^{-1}, as required. Of course, if g has finite order, then by the above considerations, g and h commute. Therefore, let g have infinite order.

Let $F\bar{G} = F(G/P)$. As we observed above, $\mathcal{U}^+(F\bar{G})$ is nilpotent. Thus, by Theorem 4.6.1, $\bar{e}$ is central in $F\bar{G}$. That is, $e^b - e \in \Delta(G, P)$ for all $b \in G$. But (using the notation above) since A is normal in G, $e^b - e \in FA$. As the product $A \times P$ is direct, $e^b - e = 0$. That is, e is central in FG.

If $(h + h^{-1})e = 0$, then the claim is trivial, so assume that $(h + h^{-1})e \neq 0$. As FHe is a field, $(h + h^{-1})e \in \mathcal{U}(FHe)$. Also, $(h + h^{-1})e$ is the image of a symmetric unit of FH under $FH \to FHe$. Indeed, if e is symmetric, then $(h + h^{-1})e + (1 - e)$ works; otherwise, use $(h + h^{-1})(e + e^*) + (1 - (e + e^*))$. Let $\eta = \hat{P}$. Then η is symmetric, central and square-zero. Thus, $1 + \eta(g + g^{-1}) \in \mathcal{U}^+(FG)$. If $\mathcal{U}^+(FG)$ satisfies $(x_1, \ldots, x_{p^m+1}) = 1$, then

$$(e + \eta(g + g^{-1})e, \underbrace{(h + h^{-1})e, \ldots, (h + h^{-1})e}_{p^m \text{ times}}) = e.$$

Thus, by Lemma 4.1.1,

$$\eta(((g + g^{-1})e)^{((h+h^{-1})e)^{p^m}} - (g + g^{-1})e) = 0.$$

As h is a p'-element, we may choose a suitably large m in such a way that $h^{p^m} = h$. Multiplying through by $(h + h^{-1})e$, we have

$$\eta((g+g^{-1})(h+h^{-1})-(h+h^{-1})(g+g^{-1}))e=0.$$

Thus, by Lemma 1.3.8,

$$[g+g^{-1},h+h^{-1}]e \in \Delta(G,P).$$

Choosing a transversal X to T in G containing g and g^{-1}, we note that every element of FG can be written uniquely in the form $\sum_{i,j,k} \alpha_i q_j g_k$, with $\alpha_i \in FA$, $q_j \in P$, $g_k \in X$. The map $FG \rightarrow F\bar{G}$ simply replaces each q_j with 1. Thus, if q_j is always 1 to begin with, a nonzero element of FG does not map to zero in $F\bar{G}$. But as A is normal in G, this is clearly true for $[g+g^{-1},h+h^{-1}]e$. That is,

$$[g+g^{-1},h+h^{-1}]e = 0.$$

It follows that

$$g(h+h^{-1}-h^g-(h^{-1})^g)e+g^{-1}(h+h^{-1}-h^{g^{-1}}-(h^{-1})^{g^{-1}})e = 0.$$

As g and g^{-1} lie in different cosets modulo T, we get

$$(h+h^{-1}-h^g-(h^{-1})^g)e = 0,$$

as claimed. We are done. □

Lemma 4.7.2. *Let F be an infinite field of characteristic $p > 2$ and G a nontorsion group with $1 < |P| < \infty$. Suppose that $\mathscr{U}^+(FG)$ is nilpotent. Let H be a finite p'-subgroup of G such that FH has a nonsymmetric primitive idempotent e. If N is the kernel of $g \mapsto ge$, then G/N is nilpotent and $(G/N)'$ is a p-group.*

Proof. As in the proof of the last lemma, we see that e is central in FG. We notice that every $\alpha e \in \mathscr{U}(FGe)$ is the image of the symmetric unit

$$\alpha e + \alpha^* e^* + (1-(e+e^*))$$

under $FG \rightarrow FGe$. In particular, since $\mathscr{U}^+(FG)$ is nilpotent, we see that $\mathscr{U}(FGe)$ is nilpotent. Thus, since G/N is embedded in $\mathscr{U}(FGe)$, we find that G/N is nilpotent.

Since N is a p'-group, G/N has a p-element; hence, it has an element zN of order p in its centre. Then z has order pk for some k relatively prime to p. Replacing z with z^k, we may assume that z has order p. Furthermore, for any $g \in G$, $(z,g) \in N \cap P = 1$. Thus, z is central. Let $\eta = \hat{z}$. Then η is central, symmetric and square-zero.

Take any $g, h \in G$. Then choosing m such that $\mathscr{U}(FGe)$ satisfies $(x_1, \ldots, x_{p^m+1}) = 1$, we have

$$(e+\eta ge, \underbrace{he, \ldots, he}_{p^m \text{ times}}) = e;$$

hence, by Lemma 4.1.1,

$$\eta(g^{h^{p^m}} - g)e = 0.$$

Therefore,

$$\eta((g,h^{p^m})-1)e = 0.$$

By Lemma 1.3.8,

$$((g,h^{p^m})-1)e \in \Delta(G,\langle z\rangle).$$

Letting $F\bar{G} = F(G/\langle z\rangle)$, we have $(\bar{g},\bar{h}^{p^m})\bar{e} = \bar{e}$. Thus, $(\bar{g},\bar{h}^{p^m})$ lies in the group generated by the support of $\bar{e}$. In particular, $(g,h^{p^m}) \in H \times \langle z\rangle$.

Let us say that $(g,h^{p^m}) = az^i$, $a \in H$, $i \geq 0$. Now, $\bar{a}\bar{e} = \bar{e}$, hence $ae - e \in \Delta(G,\langle z\rangle)$. But $ae - e \in FH$, and the product $H \times \langle z\rangle$ is direct. Thus, $ae = e$ and $a \in N$. We have

$$g^{h^{p^m}} = gaz^i,$$

hence

$$\begin{aligned}
g^{h^{2p^m}} &= g^{h^{p^m}}a^{h^{p^m}}z^i\\
&= gaz^i a^{h^{p^m}}z^i\\
&= ga'z^{2i}
\end{aligned}$$

for some $a' \in N$. After p iterations, we obtain

$$g^{h^{p^{m+1}}} = gbz^{ip} = gb$$

for some $b \in N$. That is, $(G/N)^{p^{m+1}} \subseteq \zeta(G/N)$. By Proposition 1.3.7, $(G/N)'$ is a p-group, and we are done. $\qquad\square$

We also need the following easy observation.

Lemma 4.7.3. *Let I be an ideal in a ring R. If $u_1,\ldots,u_n \in \mathscr{U}(R)$, and each $u_i \equiv 1$ (mod I), then $(u_1,\ldots,u_n) \equiv 1$ (mod I^n).*

Proof. Our proof is by induction on n. It will suffice to assume, for some $n \geq 1$, that $u \equiv 1$ (mod I) and $v \equiv 1$ (mod I^n), and show that $(u,v) \equiv 1$ (mod I^{n+1}). Let us write $u = 1+a$, $u^{-1} = 1+a'$, $v = 1+b$ and $v^{-1} = 1+b'$, where $a,a' \in I$, $b,b' \in I^n$. Then

$$(u,v) = (1+a')(1+b')(1+a)(1+b) \equiv (1+a'+b')(1+a+b) \quad (\text{mod } I^{n+1}).$$

Recalling that $a'+a+a'a = 0 = b'+b+b'b$, we see that $(u,v) \equiv 1$ (mod I^{n+1}), as required. $\qquad\square$

The second part of the main result of [69] follows.

Theorem 4.7.4. *Let F be an infinite field of characteristic $p > 2$ and G a group containing elements of infinite order. Letting P denote the set of p-elements of G, suppose further that $1 < |P| < \infty$. If $\mathscr{U}^+(FG)$ is nilpotent, then P is a (normal) subgroup of G, and G/P satisfies the conditions of Theorem 4.6.1 (for the semiprime case). In addition,*

1. *the torsion elements of G form a (normal) subgroup T, and $T = A \times P$, where A is an abelian p'-group;*

2. *if $g \in G$ and $h \in A$, then $h^g = h$ or h^{-1}; and*
3. *if A has a finite subgroup H such that FH has a nonsymmetric primitive idempotent e, then, if N is the kernel of $g \mapsto ge$, G/N is nilpotent and $(G/N)'$ is a p-group.*

Conversely, if $\mathscr{U}^+(F(G/P))$ is nilpotent, FG satisfies the three conditions above and G/T is a u.p. group, then $\mathscr{U}^+(FG)$ is nilpotent.

Proof. We have already seen the necessity, so let us verify the sufficiency. We claim that, under the given conditions, $\mathscr{U}^+(FG)$ satisfies $(x_1, \ldots, x_{2|P|}) = 1$. Suppose this is not the case, and choose $\alpha_1, \ldots, \alpha_{2|P|} \in \mathscr{U}^+(FG)$ such that $(\alpha_1, \ldots, \alpha_{2|P|}) \neq 1$. Let X be a transversal of T in G containing 1. Then for each i, write

$$\alpha_i = \sum_j \beta_{ij} q_{ij} g_{ij}$$

and

$$\alpha_i^{-1} = \sum_j \beta'_{ij} q'_{ij} g_{ij}$$

with each $\beta_{ij}, \beta'_{ij} \in FA$, each $q_{ij}, q'_{ij} \in P$ and each $g_{ij} \in X$. For all g_{ij} and g_{kl} appearing in the above expressions, write $g_{ij}g_{kl} = aqh$, with $a \in A$, $q \in P$, $h \in G$. Then let E be the (finite) subgroup of A generated by the supports of all of the β_{ij} and β'_{ij}, as well as all of the a.

Let e be a primitive idempotent of FE. Working in $F\bar{G} = F(G/P)$, we see from Theorem 4.6.1 that $\bar{e}$ is central. Thus, for any $g \in G$, $e^g - e \in \Delta(G,P)$. But $e^g - e \in FA$ as well, hence $e^g - e = 0$. That is, e is central in FG. Thus, in order to obtain a contradiction, it suffices to show that for each such e, $(\alpha_1, \ldots, \alpha_{2|P|})e = e$.

Suppose, first of all, that e is symmetric. Working modulo P, Remark 1.4.10 tells us that $\alpha_i e \equiv \lambda_i g_i \pmod{\Delta(G,P)}$, where λ_i lies in the field FEe and $g_i \in X$. Thus, $\alpha_i e = \lambda_i g_i + \rho_i$ for some $\rho_i \in \Delta(G,P)$. That is,

$$\alpha_i e = \lambda_i (1 + \delta_i) g_i,$$

where $\delta_i = \lambda_i^{-1} \rho_i g_i^{-1} \in \Delta(G,P)$. (If $\lambda_i = 0$, then $\alpha_i e \in \Delta(G,P)$; hence, by Lemma 1.1.1, $\alpha_i e$ is nilpotent. But this is surely impossible for a unit α_i; hence, $\lambda_i \in \mathscr{U}(FEe)$.)

Now, $\alpha_i e$ is symmetric, hence

$$\lambda_i (1 + \delta_i) g_i = g_i^{-1} (1 + \delta_i^*) \lambda_i^*.$$

Thus, $\lambda_i g_i \equiv g_i^{-1} \lambda_i^* \pmod{\Delta(G,P)}$. As E is a normal subgroup (since $\frac{1}{|E|}\hat{E}$ is an idempotent, hence central, as we have seen above), $g_i^{-1} \lambda_i^* = \mu_i g_i^{-1}$ for some $\mu_i \in \mathscr{U}(FEe)$. Therefore, $\lambda_i g_i \equiv \mu_i g_i^{-1} \pmod{\Delta(G,P)}$, and it follows that $\lambda_i g_i = \mu_i g_i^{-1} = g_i^{-1} \lambda_i^*$. Hence, $g_i^2 e = \lambda_i^{-1} (\lambda_i^*)^{g_i}$. Again, since E is normal, this is in FEe. But if $g_i \neq 1$, then g_i has infinite order, giving us a contradiction. Therefore, $\alpha_i e = \lambda_i (1 + \delta_i)$.

In addition, since $\lambda_i(1 + \delta_i)$ is symmetric, we have $\lambda_i \equiv \lambda_i^*$ (mod $\Delta(G,P)$). Again, since $\lambda_i - \lambda_i^* \in FE$, we see that $\lambda_i = \lambda_i^*$. Take any $g \in G$. Then for any $a \in E$, we have $a^g = a$ or a^{-1}. Thus, since λ_i is a linear combination of terms of the form $a + a^{-1}$, it follows that λ_i is central in FGe. But now

$$(\alpha_1 e, \ldots, \alpha_{|P|} e) = (\lambda_1(1 + \delta_1), \ldots, \lambda_{|P|}(1 + \delta_{|P|}))$$
$$= (1 + \delta_1, \ldots, 1 + \delta_{|P|})e,$$

which, by the preceding lemma, lies in $(1 + (\Delta(G,P))^{|P|})e$. But Lemma 1.1.1 tells us that $\Delta(G,P)$ is nilpotent; indeed, following its proof, we see that $(\Delta(G,P))^{|P|} = 0$, and this case is finished.

Suppose, on the other hand, that e is not symmetric. Then letting N be the kernel of $g \mapsto ge$, we have that G/N is nilpotent and $(G/N)'$ is a p-group. In fact, since G has only finitely many p-elements, $|(G/N)'| \leq |P|$. Now, by Lemma 4.2.4,

$$\gamma_{2|P|}(\mathcal{U}(F(G/N))) \subseteq 1 + (F(G/N))^{(2|P|)}.$$

But by Lemma 3.3.4, $(F(G/N))^{(2|P|)} = 0$. Thus, $\mathcal{U}(F(G/N))$ satisfies the identity $(x_1, \ldots, x_{2|P|}) = 1$. If $\alpha e \in \mathcal{U}(FGe)$, then $\alpha e + (1 - e) \in \mathcal{U}(FG)$. Thus, the map $\mathcal{U}(FG) \to \mathcal{U}(FGe)$ given by $\alpha \mapsto \alpha e$ is surjective. Of course, N is discarded under this map; hence, there is an epimorphism $\mathcal{U}(F(G/N)) \to \mathcal{U}(FGe)$. Therefore, $\mathcal{U}(FGe)$ satisfies $(x_1, \ldots, x_{2|P|}) = 1$, completing the proof.   □

Finally, we consider groups with infinitely many p-elements. This case certainly has the nicest answer. The third part of the result of Lee et al. [69] is the following.

Theorem 4.7.5. *Let F be an infinite field of characteristic $p > 2$ and G a nontorsion group containing infinitely many p-elements. Then the following are equivalent:*

(i) $\mathcal{U}^+(FG)$ is nilpotent;
(ii) $\mathcal{U}(FG)$ is nilpotent;
(iii) G is nilpotent and p-abelian.

In view of Theorem 4.2.1, it remains only to show that if $\mathcal{U}^+(FG)$ is nilpotent, then G is nilpotent and p-abelian. The first part of this is

Lemma 4.7.6. *Let F be an infinite field of characteristic $p > 2$ and G a nontorsion group containing infinitely many p-elements. If $\mathcal{U}^+(FG)$ is nilpotent, then G is p-abelian.*

Proof. By Theorems 2.6.5 and 2.6.11, G has a p-abelian normal subgroup A of finite index. Also, by Lemma 2.3.6, $\mathcal{U}^+(F(G/A'))$ is nilpotent. Furthermore, G is p-abelian if and only if G/A' is p-abelian. Thus, we may factor out A' and assume that A is abelian.

We know that the p-elements of G form a subgroup P. If P has bounded exponent, then by Lemma 3.4.16, G contains an infinite direct product of nontrivial finite abelian p-groups, each of which is normal in G. Thus, by Lemma 4.3.10, $(FG)^+$

is Lie nilpotent. Since, by Corollary 4.6.2, G does not contain Q_8, Theorem 3.3.1 tells us that G is p-abelian, as required. If P has unbounded exponent, then by Theorem 2.6.11, G' is a p-group of bounded exponent. If G' is finite, then we are done. Otherwise, Lemma 3.4.16 again gives us an infinite direct product of nontrivial finite normal abelian p-groups and again, we are done. $\square$

Let A be an abelian group and p any prime. We say that an element $a \in A$ has infinite p-height if, for every positive integer m, there exists $b \in A$ such that $b^{p^m} = a$. Finally, we have the

Proof of Theorem 4.7.5. In view of Lemma 4.7.6 and Theorem 4.2.1, it remains only to show that if $\mathscr{U}^+(FG)$ is nilpotent, then G is nilpotent. Let us assume that $\mathscr{U}^+(FG)$ satisfies $(x_1, \ldots, x_n) = 1$ and make some reductions.

First, if G is not nilpotent, then there exist $g_{ij} \in G$ such that $(g_{11}, g_{12}) \neq 1$, $(g_{21}, g_{22}, g_{23}) \neq 1$, and so forth. Thus, we may as well assume that G is generated by all of the g_{ij}, an element of infinite order, and a countably infinite set of p-elements. In particular, then, G is countable.

We have seen that the p-elements of G form a normal subgroup, P. By Hall's criterion (see [89, 5.2.10]), if P and G/P' are nilpotent, then G is nilpotent. But by Theorem 4.5.6, P is nilpotent and P' is finite. Thus, it suffices to show that G/P' is nilpotent. But since P' is finite, Lemma 2.3.6 tells us that $\mathscr{U}^+(F(G/P'))$ is nilpotent. We therefore replace G with G/P' and assume that P is abelian.

Let us consider the case in which no element of P (other than 1) has infinite p-height. We claim that $(FG)^+$ satisfies $[x_1, \ldots, x_n] = 0$. If not, then choose $\alpha_1, \ldots, \alpha_n \in (FG)^+$ with $[\alpha_1, \ldots, \alpha_n] \neq 0$. Let H be the subgroup of G generated by the supports of the α_i and P. Then H/P is finitely generated, say $H/P = \langle h_1 P, \ldots, h_m P \rangle$. By Lemma 4.7.6, G' is finite, hence H' is finite. In particular, H is an FC-group. Thus, each $(H : C_H(h_i)) < \infty$. Letting C be the intersection of the $C_H(h_i)$, we have $(P : C \cap P) \leq (H : C) < \infty$. In particular, $C \cap P$ is infinite. By Prüfer's theorem (see [89, 4.3.15]), a countable abelian p-group containing no nontrivial elements of infinite p-height is a direct product of cyclic groups. Now, every element of $C \cap P$ commutes with each h_i and with P, hence with H. Thus, FH contains an infinite direct product of nontrivial central cyclic p-groups. Therefore, by Lemma 4.3.10, $(FH)^+$ satisfies $[x_1, \ldots, x_n] = 0$, giving us a contradiction and proving the claim. Since $(FG)^+$ is Lie nilpotent, Theorems 3.3.1 and 3.3.6 tell us that G is nilpotent, as required.

Therefore, we may assume that P has nontrivial elements of infinite p-height. Obviously, the elements of infinite p-height form a subgroup N. We claim that N is central in G. Suppose not. Take $a \in N$, $b \in G$ such that $(a, b) \neq 1$. For any positive integer k, choose $a' \in P$ such that $(a')^{p^k} = a$. Then

$$a \neq a^b = a(a, b) = a(a', b)^{p^k}.$$

Thus, $(a', b)^{p^k} \neq 1$. Since G' is a p-group, it follows that G' has elements of arbitrarily large order, contradicting the fact that G' is finite. Therefore, N is central in G.

Let $F\bar{G} = F(G/N)$. We claim that $\mathscr{U}^+(F\bar{G})$ satisfies $(x_1,\dots,x_n) = 1$. If not, then choose $\bar{\beta}_1,\dots,\bar{\beta}_n \in \mathscr{U}^+(F\bar{G})$ such that $(\bar{\beta}_1,\dots,\bar{\beta}_n) \neq 1$. Each $\bar{\beta}_i$ is the image under $FG \to F\bar{G}$ of some $\beta_i \in (FG)^+$. Similarly, letting $\bar{\rho}_i = (\bar{\beta}_i)^{-1}$, we can lift each $\bar{\rho}_i$ to $\rho_i \in (FG)^+$. Thus, $\bar{\beta}_i\bar{\rho}_i - 1 = 0 = \bar{\rho}_i\bar{\beta}_i - 1$, hence $\beta_i\rho_i - 1, \rho_i\beta_i - 1 \in \Delta(G,N)$. Now N is a central p-group; thus, there exists a finite subgroup N_1 of N such that all $\beta_i\rho_i - 1, \rho_i\beta_i - 1 \in \Delta(G,N_1)$. By Lemma 1.1.1, $\Delta(G,N_1)$ is nilpotent. Let us say that $(\Delta(G,N_1))^{p^t} = 0$. Then $(\beta_i\rho_i)^{p^t} = (\rho_i\beta_i)^{p^t} = 1$, hence $\beta_i \in \mathscr{U}^+(FG)$. It follows that $(\beta_1,\dots,\beta_n) = 1$, hence $(\bar{\beta}_1,\dots,\bar{\beta}_n) = 1$, and the claim is proved.

We also observe that P/N has no nontrivial elements of infinite p-height. Indeed, if bN has infinite p-height, then for any positive integer k, choose $aN \in P/N$ such that $(aN)^{p^k} = bN$. Then $a^{p^k}c = b$ for some $c \in N$. Now $c = (c')^{p^k}$ for some $c' \in P$. Thus, $(ac')^{p^k} = b$ and b has infinite p-height, hence $b \in N$. If P/N is infinite, then since $\mathscr{U}^+(F(G/N))$ satisfies $(x_1,\dots,x_n) = 1$, we see from the case where P has no nontrivial elements of infinite p-height that G/N is nilpotent. Since N is central, G is nilpotent. Thus, we may assume that P/N is finite. Since P has nontrivial elements of infinite p-height, it certainly has unbounded exponent. Thus, N has unbounded exponent.

Let a_1 and a_2 be nonidentity elements of N, and take any $g,h \in G$. Suppose a_1 has order p^r. Since a_1 is central, $1 + (a_1 - a_1^{-1})(g - g^{-1})$ is symmetric and

$$(1 + (a_1 - a_1^{-1})(g - g^{-1}))^{p^r} = 1 + (a_1^{p^r} - a_1^{-p^r})(g^{p^r} - g^{-p^r}) = 1.$$

Thus, $1 + (a_1 - a_1^{-1})(g - g^{-1}) \in \mathscr{U}^+(FG)$. Let $\eta = \hat{a}_2$. Then η is symmetric, central and square-zero, hence $1 + \eta(h + h^{-1}) \in \mathscr{U}^+(FG)$. Choosing t so that $p^t + 1 \geq n$, we have

$$(1 + \eta(h + h^{-1}), \underbrace{1 + (a_1 - a_1^{-1})(g - g^{-1}), \dots, 1 + (a_1 - a_1^{-1})(g - g^{-1})}_{p^t \text{ times}}) = 1.$$

Thus, by Lemma 4.1.1,

$$\eta((h + h^{-1})^{(1 + (a_1 - a_1^{-1})(g - g^{-1}))^{p^t}} - (h + h^{-1})) = 0.$$

Multiplying through by $(1 + (a_1 - a_1^{-1})(g - g^{-1}))^{p^t}$, we get

$$\eta[h + h^{-1}, 1 + (a_1^{p^t} - a_1^{-p^t})(g^{p^t} - g^{-p^t})] = 0.$$

Discarding the 1, pulling out $a_1^{p^t} - a_1^{-p^t}$ and multiplying by $a_1^{p^t}$, we get

$$(a_1^{2p^t} - 1)\eta[h + h^{-1}, g^{p^t} - g^{-p^t}] = 0.$$

But this equation holds for any $a_1 \in N$. Since N has unbounded exponent, there are infinitely many different elements $a_1^{2p^t}$. Thus, by Lemma 1.5.8,

$$\eta[h+h^{-1}, g^{p^t} - g^{-p^t}] = 0.$$

By Lemma 1.3.8,

$$[h+h^{-1}, g^{p^t} - g^{-p^t}] \in \Delta(G, \langle a_2 \rangle)$$

for all $g, h \in G$. That is, in $F(G/\langle a_2 \rangle)$, all elements of $G/\langle a_2 \rangle$ satisfy

$$[h+h^{-1}, g^{p^t} - g^{-p^t}] = 0.$$

As $\langle a_2 \rangle$ is central, it suffices to show that $G/\langle a_2 \rangle$ is nilpotent. By Lemma 2.3.6, $\mathscr{U}^+(F(G/\langle a_2 \rangle))$ is nilpotent. Thus, we replace G with $G/\langle a_2 \rangle$ and assume that the elements of G satisfy the above identity.

Increase t if necessary so that $|P/N| \le p^t$. We claim that for every $g, h \in G$, we have $(g^{p^t}, h) = 1$. Suppose, first of all, that g is torsion. We know that the torsion elements of G form a subgroup $T = A \times P$, where A is an abelian p'-group. Then $g^{p^t} \in A \times N$, where N is central. Thus, $(h, g^{p^t}) \in A \cap P$, since G' is a p-group. But $A \cap P = 1$, and the claim holds here.

So, suppose that g has infinite order. Expanding $[h + h^{-1}, g^{p^t} - g^{-p^t}] = 0$, we get

$$hg^{p^t} + h^{-1}g^{p^t} + g^{-p^t}h + g^{-p^t}h^{-1} = hg^{-p^t} + h^{-1}g^{-p^t} + g^{p^t}h + g^{p^t}h^{-1}.$$

It is clear that if $p > 3$, then $hg^{p^t} \in \{hg^{-p^t}, h^{-1}g^{-p^t}, g^{p^t}h, g^{p^t}h^{-1}\}$. If $p = 3$, we could conceivably have precisely three of the elements on the left-hand side agreeing. But it is trivial to check that if three of them agree, then all four agree, so in any case, $hg^{p^t} \in \{hg^{-p^t}, h^{-1}g^{-p^t}, g^{p^t}h, g^{p^t}h^{-1}\}$.

In fact, the case $hg^{p^t} = hg^{-p^t}$ cannot occur, since g is torsion in this case. Suppose that h is torsion. If $hg^{p^t} = h^{-1}g^{-p^t}$, then $g^{2p^t} = h^{-2} \in T$, giving us a contradiction. The remaining two cases show that $h^{g^{p^t}} = h$ or h^{-1}. Suppose that there exists a torsion element h such that $h^{g^{p^t}} = h^{-1}$. Take $z \in N$ of order p. Then hz is also torsion, hence $(hz)^{g^{p^t}} = hz$ or $(hz)^{-1}$. But z is also central, hence $(hz)^{g^{p^t}} = h^{-1}z$. Thus, either $hz = h^{-1}z$ (in which case $h = h^{-1}$ and $(g^{p^t}, h) = 1$) or $h^{-1}z^{-1} = h^{-1}z$ (in which case $z^2 = 1$, which is impossible). Therefore, $(g^{p^t}, h) = 1$ for all $g \in G$ and all $h \in T$.

Suppose, on the other hand, that h has infinite order. If $hg^{p^t} = h^{-1}g^{-p^t}$, then $(hg^{p^t})^2 = (h, g^{p^t}) \in G' \le P$. Thus, hg^{p^t} is torsion. As we saw above, g^{p^t} commutes with hg^{p^t}, hence with h. Thus, we have $h^{g^{p^t}} = h$ or h^{-1} and, as above, we get $(g^{p^t}, h) = 1$, proving the claim.

Thus, $G/\zeta(G)$ is a p-group. Since G' is finite, Lemma 4.3.12 shows that $G/\zeta(G)$ is nilpotent. Hence, G is nilpotent, and the theorem is proved. $\square$

Chapter 5
The Bounded Engel Property

5.1 Introduction

Weakening the condition that $\mathscr{U}(FG)$ is nilpotent slightly, we ask instead when it is bounded Engel; that is, when does $\mathscr{U}(FG)$ satisfy

$$(x_1, \underbrace{x_2, \ldots, x_2}_{n \text{ times}}) = 1$$

for some n?

Bounded Engel groups are not quite as well understood as nilpotent groups. Certainly, the most famous result, due to Zorn, is that every finite bounded Engel group is nilpotent (see [89, 12.3.4]).

If char $F = 0$ or char $F = p > 0$ and G has no p-elements, then the solution was found by Bovdi and Khripta in [14] and [15]. They also presented solutions for other special cases. Subsequently, Riley solved the problem for torsion groups in [87]. The general result, showing that if G has a p-element, and $\mathscr{U}(FG)$ is bounded Engel, then FG is bounded Lie Engel, was presented in Bovdi [11]. The converse had already been established in a much more general setting by Shalev in [97].

We also consider the symmetric units. In order to make use of the results in Chapter 2, F must be an infinite field of characteristic different from 2. Unfortunately, we do not have an analogue of Lemma 4.1.2 with which to work. That is, let H be a subset of any group. To the best of the author's knowledge, it is not presently known if, whenever H satisfies the group identity

$$(x_1, \underbrace{x_2, \ldots, x_2}_{n \text{ times}}) = 1,$$

we must have $\langle H \rangle$ satisfying the same identity. It would, in particular, be desirable to show that if $\mathscr{U}^+(FG)$ is bounded Engel, then so is $\langle \mathscr{U}^+(FG) \rangle$, but it is not presently known if this is true.

G.T. Lee, *Group Identities on Units and Symmetric Units of Group Rings*,
Algebra and Applications 12, DOI 10.1007/978-1-84996-504-0_5,
© Springer-Verlag London Limited 2010

Let char $F = p > 2$ and let G be a finite group. Then if we assume that $\langle \mathscr{U}^+(FG) \rangle$ is bounded Engel, it follows that $\langle \mathscr{U}^+(\mathbb{Z}_p G) \rangle$ is bounded Engel. As the latter is a finite group, from the theorem of Zorn we conclude that $\mathscr{U}^+(\mathbb{Z}_p G)$ is nilpotent. Fortunately, the author's results in Chapter 4 for torsion groups did not depend upon the size of the field, so these classifications are available to us. In particular, we can use the fact that symmetric units of relatively prime order must commute, and so forth.

The first results concerning the symmetric units were proved recently by Lee and Spinelli in [74]. Here, the torsion groups G such that $\langle \mathscr{U}^+(FG) \rangle$ is bounded Engel were classified. If G is not torsion, then the problem remains open in general, but we can provide the answer (for either $\mathscr{U}^+(FG)$ or the subgroup it generates) if FG is semiprime.

We will present the results for $\mathscr{U}(FG)$ in the next section, and those for $\mathscr{U}^+(FG)$ in the final section.

5.2 Bounded Engel Unit Groups

Let F be a field and G a group. If G has no elements of order divisible by char F, then the solution to the bounded Engel problem is very similar to Theorem 4.2.9. Indeed, the same exceptional case occurs here, and we simplify matters by assuming that F is infinite, referring the reader to Bovdi and Khripta [15, Theorem 1.3] for the finite field case.

Theorem 5.2.1. *Let F be a field of characteristic $p \geq 0$ and G a group such that, if $p > 0$, then G has no p-elements. If $\mathscr{U}(FG)$ is bounded Engel, then G is bounded Engel, the torsion elements of G form a (normal) subgroup T, and either*

1. *T is central in G; or*
2. *$|F| = p$ is a Mersenne prime, T is an abelian group of exponent dividing $p^2 - 1$, and for every $h \in T$, $g \in G$, we have $h^g = h$ or h^p.*

Conversely, if G is a bounded Engel group, G/T is a u.p. group, and G satisfies one of the two conditions above, then $\mathscr{U}(FG)$ is bounded Engel.

Proof. Suppose that $\mathscr{U}(FG)$ is bounded Engel. Let us dispense with the case in which G is torsion first. If $p = 0$, then by Corollary 1.2.21, G is abelian. Otherwise, we may as well assume that $F = \mathbb{Z}_p$. By Theorem 1.2.27, FG satisfies a polynomial identity, hence, by Proposition 1.1.4, G is locally finite. Thus, it suffices to assume that G is finite, hence $\mathscr{U}(\mathbb{Z}_p G)$ is finite. By Zorn's theorem, $\mathscr{U}(\mathbb{Z}_p G)$ is nilpotent, and by Proposition 4.2.6, G is abelian.

Thus, let G have elements of infinite order. We assume here that F is infinite. By Theorem 1.4.9, T is an abelian group. That G must be bounded Engel is obvious. In order to show that T is central in G, it suffices to assume that $G = \langle g, h \rangle$, where $g \in T$ and h has infinite order. By Theorem 1.4.9, the idempotent $\frac{1}{o(g)}\hat{g}$ is central, hence $\langle g \rangle$ is a normal subgroup. Thus, $T = \langle g \rangle$ is finite. We now apply the proof of Theorem 4.2.9 verbatim to show that T is central in G.

Conversely, suppose that G is n-Engel, G/T is a u.p. group, and T is central. We claim that $\mathscr{U}(FG)$ is n-Engel. Take any $\alpha, \beta \in \mathscr{U}(FG)$. By Remark 1.4.10, there exists a finite subgroup E of T such that for every primitive idempotent e of FE, $\alpha e = \lambda_1 g_1$ and $\beta e = \lambda_2 g_2$ for some $\lambda_i \in \mathscr{U}(FEe)$, and $g_i \in G$. Thus,

$$(\alpha, \underbrace{\beta, \ldots, \beta}_{n \text{ times}})e = (\lambda_1 g_1, \underbrace{\lambda_2 g_2, \ldots, \lambda_2 g_2}_{n \text{ times}}) = (g_1, \underbrace{g_2, \ldots, g_2}_{n \text{ times}})e,$$

since each λ_i is central. But G is n-Engel, hence this is e. Summing over all e, we get our required identity. $\square$

The remaining case is summed up in the following theorem.

Theorem 5.2.2. *Let F be a field of characteristic $p > 0$ and G a group having a p-element. Then $\mathscr{U}(FG)$ is bounded Engel if and only if G is nilpotent and G has a p-abelian normal subgroup of finite p-power index.*

Let us first consider the necessity part of the theorem, which is due to Riley [87] (for torsion groups) and Bovdi [11] (in general). We need a couple of group-theoretic lemmas from [15].

Lemma 5.2.3. *Let G be a bounded Engel group. If G has a nilpotent normal subgroup N such that G/N is cyclic, then G is nilpotent.*

Proof. Suppose that G is n-Engel. Let $G/N = \langle Ng \rangle$. Fix a positive integer m and, as usual, write $\gamma_m(N)$ for the mth term of the lower central series of N. We observe that if $a \in \gamma_m(N)$ and for each i, $1 \leq i \leq n$, we have either $b_i \in N$ or $b_i = g$, then $(a, b_1, \ldots, b_n) \in \gamma_{m+1}(N)$. Indeed, this is obvious from the definition if any $b_i \in N$, and if not, then we are dealing with

$$(a, \underbrace{g, \ldots, g}_{n \text{ times}}) = 1,$$

since G is n-Engel.

But now if we have any sequence c_i, with each c_i either equal to g or lying in N, then since G/N is abelian, we have $(c_1, c_2) \in N = \gamma_1(N)$. Hence, by our observation above,

$$(c_1, c_2, c_3, \ldots, c_{n+2}) \in \gamma_2(N),$$

$$(c_1, c_2, c_3, \ldots, c_{2n+2}) \in \gamma_3(N),$$

and so forth. As N is nilpotent, there exists a k such that $(c_1, \ldots, c_k) = 1$ for all $c_i \in N \cup \{g\}$. But $G = \langle N, g \rangle$. Thus, by Lemma 4.1.2, G is nilpotent. $\square$

Lemma 5.2.4. *Let G be a bounded Engel group. Suppose that G has a normal subgroup N, and either*

1. N is nilpotent, and G/N is finitely generated nilpotent, or
2. N is finite and G/N is nilpotent.

Then G is nilpotent.

Proof. Let us prove the first part by induction on the nilpotency class of G/N. Let $H/N = \zeta(G/N)$. By our inductive hypothesis, it suffices to show that H is nilpotent, since G/H has smaller nilpotency class than G/N. But H/N is abelian and, as every subgroup of a finitely generated nilpotent group is finitely generated, we know that H/N is finitely generated. Therefore it is a direct product of cyclic groups. But now we apply the preceding lemma repeatedly in order to obtain the first part of our result.

For the second part, we note that N is nilpotent, being a finite bounded Engel group. Let

$$1 = N_0 \leq N_1 \leq \cdots \leq N_k = N$$

be a central series for N. Of course, $N \leq \phi(G)$, so $C_G(N)$ is a subgroup of finite index in G. Let $H = NC_G(N)$. Then H is also a normal subgroup of finite index in G. As H/N is nilpotent, let

$$N/N = H_0/N \leq H_1/N \leq \cdots \leq H_m/N = H/N$$

be a central series for H/N. We observe, then, that

$$1 = N_0 \leq N_1 \leq \cdots \leq N \leq H_1 \leq H_2 \leq \cdots \leq H$$

is a central series for H. Thus, H is nilpotent. But G/H is a finite nilpotent group, hence, by the first part of the lemma, G is nilpotent.   □

We can now prove the necessity portion of Theorem 5.2.2.

Proposition 5.2.5. *Let F be a field of characteristic $p > 0$ and G a group having p-elements. If $\mathcal{U}(FG)$ is bounded Engel, then G is nilpotent, and G has a p-abelian normal subgroup of finite p-power index.*

Proof. Suppose either that G is torsion or G has infinitely many p-elements. Then by Theorems 1.2.27, 1.5.10 and 1.5.16, G has a p-abelian normal subgroup A of finite index. As A is bounded Engel and finite-by-abelian, the previous lemma tells us that A is nilpotent. Also, G/A is finite and bounded Engel, hence nilpotent, and applying the previous lemma a second time shows that G is nilpotent. Thus, since G has a p-element, it has one in its centre. By Lemma 4.2.2, $G/\zeta(G)$ is a p-group of bounded exponent. Thus, $A\zeta(G)$ is a p-abelian subgroup of finite p-power index, and we are done.

We may, therefore, assume that G is nontorsion and has only a finite number of p-elements. In view of Theorem 1.5.6, the p-elements of G form a subgroup, P. Let $\eta = \hat{P}$. Then η is central and square-zero. If $\mathcal{U}(FG)$ is p^m-Engel, then by Lemma 4.1.1, we get

$$1 = (1 + \eta g, \underbrace{h, \ldots, h}_{p^m \text{ times}}) = 1 + \eta(g^{h^{p^m}} - g)$$

for any $g, h \in G$. Thus, $(g, h^{p^m})\eta = \eta$, hence $(g, h^{p^m}) \in P$.

If we can show that $G' \leq P$, then we will be done, since the previous lemma then shows that G is nilpotent. Thus, let us factor out P and simply assume that G is a p'-group in which

$$(g, \underbrace{h, \ldots, h}_{p^m \text{ times}}) = (g, h^{p^m}) = 1$$

for all $g, h \in G$. Our aim is to show that G is abelian.

Fix $g, h \in G$ such that $gh \neq hg$, and let k be the smallest positive integer such that

$$(g, \underbrace{h, \ldots, h}_{k \text{ times}}) = 1.$$

(We know that $2 \leq k \leq p^m$.) Replacing g with

$$(g, \underbrace{h, \ldots, h}_{k-2 \text{ times}})$$

if necessary, let us assume that $(g, h) \neq 1$ but $(g, h, h) = 1$. Since h commutes with (g, h), we get $(g, h)^2 = g^{-1}h^{-1}g(g, h)h = (g, h^2)$ and similarly, by induction, $(g, h)^{p^m} = (g, h^{p^m}) = 1$. But G has no p-elements, so $(g, h) = 1$, giving a contradiction and completing the proof. □

Taking into account Theorem 3.1.2, the proof of Theorem 5.2.2 will be complete once we have proved the following result from Shalev [97].

Proposition 5.2.6. *Let F be a field of characteristic different from zero and R an F-algebra. If R is bounded Lie Engel, then $\mathscr{U}(R)$ is bounded Engel.*

In fact, if char $F = 0$, then combining the results of Shalev [97] with those of Zel'manov [103], it is possible to reach an even stronger conclusion; namely, that $\mathscr{U}(R)$ is nilpotent. We will not need that fact, however.

We do need to borrow the following classical result of Gruenberg.

Proposition 5.2.7. *Every finitely generated, solvable Lie algebra is nilpotent.*

Proof. See [46, Theorem 1']. □

The other result of [97] is also of interest.

Proposition 5.2.8. *Let F be a field of characteristic $p > 0$ and R a finitely generated F-algebra. If R is bounded Lie Engel, then R is Lie nilpotent.*

Proof. We know that $R/J(R)$ is semiprimitive. Let us say that $R/J(R)$ is the subdirect product of the primitive algebras R_i, $i \in I$. Then as each R_i is a homomorphic image of R, it is bounded Lie Engel. However, by Proposition 1.5.11, R_i satisfies the same polynomial identities as $M_{n_i}(F_i')$, where n_i is a positive integer and F_i' is the centre of R_i. But noting by induction that

$$\left[\begin{pmatrix} 0 & 0 \\ 1 & 0 \end{pmatrix}, \underbrace{\begin{pmatrix} 1 & 0 \\ 0 & 0 \end{pmatrix}, \ldots, \begin{pmatrix} 1 & 0 \\ 0 & 0 \end{pmatrix}}_{k \text{ times}} \right] = \begin{pmatrix} 0 & 0 \\ 1 & 0 \end{pmatrix}$$

for any positive integer k, we see that $M_{n_i}(F_i')$ is not bounded Lie Engel unless $n_i = 1$. Thus, $M_{n_i}(F_i')$, and hence R_i, is commutative. As $R/J(R)$ is embedded in $\prod_{i \in I} R_i$, we conclude that $R/J(R)$ is commutative. In particular, $[R,R] \subseteq J(R)$.

By Proposition 1.5.12, $J(R)$ is nilpotent. Of course, for any ideal I, $[I,I] \subseteq I^2$, hence $J(R)$ is Lie solvable. As $[R,R] \subseteq J(R)$, we see that R is Lie solvable.

Furthermore, if $a, b \in R$, and R is Lie p^m-Engel, then

$$0 = [a, \underbrace{b, \ldots, b}_{p^m \text{ times}}] = [a, b^{p^m}].$$

That is, b^{p^m} is central for all $b \in R$, so R is integral over Z, the centre of R. But by Shirshov's theorem (see, for instance, [91, Corollary 4.2.9]), since R is a finitely generated algebra satisfying a polynomial identity, we find that R is finitely generated as a Z-module.

Let $R = \sum_{i=1}^{t} Z r_i$, and let L be the Lie algebra generated by the r_i. Then as we know that L is solvable, the preceding proposition tells us that L is nilpotent. But, of course, $R = ZL$, and as Z is central, R is Lie nilpotent, too. We are done.   □

We are also going to need a result of Gupta and Levin [47]. Let R be any ring. Let $L_1(R) = R$, and for any $n \geq 2$, let $L_n(R)$ be the (associative) ideal of R generated by all of the elements $[a_1, a_2, \ldots, a_n]$ for all $a_i \in R$. We then define $P_n(R)$ to be the set of all $a \in L_n(R)$ such that $[a, a_1, \ldots, a_k] \in L_{n+k}(R)$ for all $k \geq 1$ and all $a_i \in R$. We shall prove

Proposition 5.2.9. *Let R be a ring and $u_1, \ldots, u_n \in \mathcal{U}(R)$ for some $n \geq 2$. Then $(u_1, \ldots, u_n) \in 1 + P_n(R)$.*

We begin the proof with

Lemma 5.2.10. *Let R be a ring and n a positive integer. Then $P_n(R)$ is an additive subgroup of R and furthermore, if $a \in L_n(R)$, then $a \in P_n(R)$ if and only if $[a,b] \in P_{n+1}(R)$ for all $b \in R$.*

Proof. This is clear from the definition of $P_n(R)$.   □

Lemma 5.2.11. *Let R be a ring and n a positive integer. If $a \in P_n(R)$ and $u \in \mathcal{U}(R)$, then $u^{-1}[a,u] \in P_{n+1}(R)$.*

Proof. Let k be a positive integer, and take any $b_1, \ldots, b_k \in R$. We claim that

$$[u^{-1}[a,u], b_1, \ldots, b_k]$$

is a $\mathbb{Z}$-linear combination of terms of the form

$$u^{-1}[a, c_1, \ldots, c_k, u]$$

and terms of the form

$$u^{-1}[a, c_1, \ldots, c_{k+1}]u$$

for various $c_i \in R$. Since $a \in P_n(R)$, all of these terms lie in $L_{n+k+1}(R)$; hence, by definition, we will have $u^{-1}[a,u] \in P_{n+1}(R)$, and the proof will be complete.

We proceed by induction on k. If $k = 1$, then we observe that

$$[u^{-1}[a,u], b_1] = u^{-1}[a, b_1, u] + u^{-1}[a, u, b_1 u^{-1}]u - u^{-1}[a, b_1 u^{-1}, u]u,$$

which is of the correct form.

Assume that our claim is true for k, and consider the $k+1$ case. By our inductive hypothesis,

$$[u^{-1}[a,u], b_1, \ldots, b_k, b_{k+1}]$$

is a $\mathbb{Z}$-linear combination of terms of the form

$$[u^{-1}[a, c_1, \ldots, c_k, u], b_{k+1}]$$

and

$$[u^{-1}[a, c_1, \ldots, c_{k+1}]u, b_{k+1}]$$

for various c_i. But the first of these terms can be seen to be of the correct form by applying the $k = 1$ case with $[a, c_1, \ldots, c_k]$ in place of a and b_{k+1} in place of b_1. The second of these terms is also of the correct form, since

$$[u^{-1}[a, c_1, \ldots, c_{k+1}]u, b_{k+1}] = u^{-1}[a, c_1, \ldots, c_{k+1}, u b_{k+1} u^{-1}]u.$$

We are done.   □

Lemma 5.2.12. *In any ring R, we have*

$$[abc, d] = [ac, db] - [c, adb] + [c, a, db] + [bc, ad] - [bc, a, d].$$

Proof. Routine.   □

Now, we have the

Proof of Proposition 5.2.9. Noting that $P_1(R) = R$, it suffices, by induction, to assume that $u, v \in \mathscr{U}(R)$ and $u - 1 \in P_n(R)$, and show that $(u,v) - 1 \in P_{n+1}(R)$. Now, $[u,v] = [u-1, v] \in L_{n+1}(R)$, since $u - 1 \in P_n(R)$. Hence,

$$(u,v) - 1 = u^{-1}v^{-1}[u,v] \in L_{n+1}(R).$$

Thus, in view of Lemma 5.2.10, it suffices to show that

$$[u^{-1}v^{-1}[u,v],b] \in P_{n+2}(R)$$

for all $b \in R$.

Now, by the last lemma,

$$[u^{-1}v^{-1}[u,v],b] = [u^{-1}[u,v],bv^{-1}] - [u,v,u^{-1}bv^{-1}] + [u,v,u^{-1},bv^{-1}]$$
$$+ [v^{-1}[u,v],u^{-1}b] - [v^{-1}[u,v],u^{-1},b].$$

Let us consider each of these terms. But

$$[u^{-1}[u,v],bv^{-1}] = [u,u^{-1}v,bv^{-1}] = [u-1,u^{-1}v,bv^{-1}] \in P_{n+2}(R),$$

by Lemma 5.2.10. Similarly,

$$[u,v,u^{-1}bv^{-1}] = [u-1,v,u^{-1}bv^{-1}] \in P_{n+2}(R)$$

and

$$[u,v,u^{-1},bv^{-1}] = [u-1,v,u^{-1},bv^{-1}] \in P_{n+3}(R) \subseteq P_{n+2}(R).$$

By Lemma 5.2.11, $v^{-1}[u,v] = v^{-1}[u-1,v] \in P_{n+1}(R)$, so the remaining terms also lie in $P_{n+2}(R)$. We are done.   □

Let us now wrap up the proof of Proposition 5.2.6, and hence of Theorem 5.2.2. Suppose that R is Lie n-Engel. Let B be the relatively free F-algebra of rank 4 determined by the identity

$$[x_1,\underbrace{x_2,\ldots,x_2}_{n \text{ times}}].$$

By Proposition 5.2.8, B is Lie nilpotent. Let us say that B satisfies

$$[x_1,x_2,\ldots,x_m] = 0.$$

Take any $u,v \in \mathscr{U}(R)$ and let S be the F-algebra generated by u, u^{-1}, v and v^{-1}. Then S is a homomorphic image of B, hence S also satisfies $[x_1,\ldots,x_m] = 0$.

But in view of Proposition 5.2.9, we have

$$(u,\underbrace{v,\ldots,v}_{m-1 \text{ times}}) - 1 \in P_m(S) \subseteq L_m(S),$$

and as S satisfies $[x_1,\ldots,x_m] = 0$, we have $L_m(S) = 0$. Bearing in mind that m is independent of the choice of u and v, we conclude that $\mathscr{U}(R)$ is $(m-1)$-Engel. The proof is complete.

Comparing Theorems 5.2.1 and 5.2.2 with Theorem 3.1.2, we record

Corollary 5.2.13. *Let F be a field and G a torsion group. Then $\mathscr{U}(FG)$ is bounded Engel if and only if FG is bounded Lie Engel.*

5.3 Bounded Engel Symmetric Units

We now classify the torsion groups G such that $\langle \mathscr{U}^+(FG) \rangle$ is bounded Engel, when F is an infinite field of characteristic different from 2. All of the results in this section are from Lee and Spinelli [74]. Once again, the problem breaks down into two cases: those groups that contain the quaternions and those that do not. We consider the latter class first.

Theorem 5.3.1. Let F be an infinite field of characteristic $p \neq 2$ and G a torsion group not containing Q_8. Then the following are equivalent:

(i) $\langle \mathscr{U}^+(FG) \rangle$ is bounded Engel;
(ii) $\mathscr{U}(FG)$ is bounded Engel;
(iii) either $p = 0$ and G is abelian, or $p > 2$, G is nilpotent and G has a p-abelian normal subgroup of finite p-power index.

Proof. Of course, it is obvious that (ii) implies (i). If (iii) holds, then by Theorem 3.1.2, FG is bounded Lie Engel. Thus, by Proposition 5.2.6, (ii) holds. Therefore, it suffices to show that (i) implies (iii).

Suppose that $\langle \mathscr{U}^+(FG) \rangle$ is bounded Engel. If $p = 0$, then Proposition 2.3.11 tells us that G is abelian. Therefore, assume that $p > 2$. We claim that $G = H \times P$, where H is an abelian p'-group and P is a p-group. By Proposition 2.4.3, G is locally finite. Thus, in order to prove the claim, it is sufficient to assume that G is finite. But then $\langle \mathscr{U}^+(\mathbb{Z}_p G) \rangle$ is a finite bounded Engel group, and it is therefore nilpotent. Proposition 4.3.2 then proves the claim.

Thus, it suffices to assume that G is a p-group. If FG is a prime ring, then by Proposition 4.4.1, G is the trivial group, and we are done. Therefore, in view of Proposition 4.3.1, let us assume that G has a nontrivial finite normal subgroup N. But now following the proof of Proposition 4.3.2, we conclude that $(F(G/N))^+$ is bounded Lie Engel. Thus, by Theorem 3.2.8, G/N is nilpotent and has a p-abelian normal subgroup A/N of finite index. It follows immediately that A is of finite index in G, and as $(A/N)' = A'N/N$, we see that A is p-abelian. Hence, it remains only to show that G is nilpotent.

Now, N/N' is a finite normal subgroup of G/N', so it is surely contained in $\phi(G/N')$. Thus, its centralizer has finite index in G/N'. By Lemma 3.2.6, we have

$$(N/N', G/N', G/N', \ldots, G/N') = 1.$$

That is,

$$(N, G, G, \ldots, G) \leq N'.$$

But by the same argument,

$$(N'/N'', G/N'', \ldots, G/N'') = 1,$$

and hence

$$(N', G, G, \ldots, G) \leq N''.$$

Therefore,

$$(N,G,G,\ldots,G) \leq N''.$$

Repeating this argument, and noting that since N is a finite p-group, it is nilpotent, we obtain

$$(N,G,G,\ldots,G) = 1.$$

But G/N is nilpotent as well, hence,

$$(G,G,\ldots,G) \leq N,$$

and therefore,

$$(G,G,\ldots,G) = 1,$$

as required.   □

Of course, the solution is different if G contains the quaternions.

Theorem 5.3.2. *Let F be an infinite field of characteristic $p \geq 0$ and G a torsion group containing Q_8. Then $\langle \mathscr{U}^+(FG) \rangle$ is bounded Engel if and only if either*

1. $p = 0$ and $G \simeq Q_8 \times E$, where E is an elementary abelian 2-group; or
2. $p > 2$ and $G \simeq Q_8 \times E \times P$, where E is an elementary abelian 2-group and P is a p-group of bounded exponent containing a p-abelian normal subgroup of finite index.

Proof. Let us prove the necessity. The $p = 0$ case is handled by Proposition 2.3.11, so assume that $p > 2$. As in the proof of the previous theorem, we may consider finite subgroups of G. In view of Propositions 4.3.2 and 4.3.3, each finite subgroup of G is a direct product of the form $H \times P$, where H is a p'-group that is either abelian or a Hamiltonian 2-group, and P is a p-group. Thus, G has this form as well, and since G contains Q_8, we can only have $G \simeq Q_8 \times E \times P$, where E is an elementary abelian 2-group and P is a p-group. By Lemma 2.3.7, P has bounded exponent, and as $\langle \mathscr{U}^+(FP) \rangle$ is bounded Engel, the previous theorem completes the proof of the necessity.

Now let us consider the sufficiency. By Lemma 2.1.1, if $G = Q_8 \times E$, then the symmetric units of FG are central, so the characteristic zero case is complete. Assume that $p > 2$, and $G = Q_8 \times E \times P$ as in the statement of the theorem. Let A be the p-abelian normal subgroup of finite index in P. Then $E \times A$ is p-abelian and has finite index in G. Thus, by Proposition 1.1.4, FG satisfies a polynomial identity. By Lemma 1.3.14, $\Delta(G,P)$ is a nil ideal of bounded exponent. Let us say that its exponent is at most p^m.

Notice that every symmetric element of FG is a linear combination of terms of the form $gc + g^{-1}c^{-1}$, with $g \in Q_8 \times E$ and $c \in P$. But

$$gc + g^{-1}c^{-1} = (g + g^{-1}) + (g(c-1) + g^{-1}(c^{-1} - 1)).$$

Now, $g + g^{-1}$ lies in $(F(Q_8 \times E))^+$, and is therefore central, and $g(c-1) + g^{-1}(c^{-1} - 1) \in \Delta(G,P)$. Thus, every $\alpha \in (FG)^+$ is of the form $\alpha = \gamma + \delta$, with

γ central and $\delta \in \Delta(G,P)$. Now, $\alpha^{p^m} = \gamma^{p^m}$. Thus, if α is a unit, then γ is a central unit, and $\alpha = \gamma(1 + \delta')$, for some $\delta' \in \Delta(G,P)$. Clearly, every product of symmetric units must also have this form, hence every element of $\langle \mathscr{U}^+(FG) \rangle$ may be written in this way.

As we are interested only in commutators, a central unit is irrelevant; it will vanish. Thus, we may assume that every unit in which we are interested lies in the group $1 + \Delta(G,P)$. In particular, it is a unit in the F-algebra $R = F + \Delta(G,P)$. We claim that R is Lie p^m-Engel. Take any $r_i = \lambda_i + \delta_i$, $\lambda_i \in F$, $\delta_i \in \Delta(G,P)$. Then

$$[r_1, \underbrace{r_2, \ldots, r_2}_{p^m \text{ times}}] = [r_1, r_2^{p^m}] = [r_1, \lambda_2^{p^m}] = 0,$$

since λ_2 is central. The claim is proved. But then by Proposition 5.2.6, $\mathscr{U}(R)$ is bounded Engel. The proof is complete. $\qquad\qquad\qquad\qquad\qquad\qquad\qquad\square$

Comparing these last two theorems with Theorems 3.2.8 and 3.2.18, we obtain

Corollary 5.3.3. *Let F be an infinite field of characteristic different from 2 and G a torsion group. Then $\langle \mathscr{U}^+(FG) \rangle$ is bounded Engel if and only if $(FG)^+$ is bounded Lie Engel.*

We conjecture that this corollary is still true if we assume instead that $\mathscr{U}^+(FG)$ is bounded Engel, but that problem is currently open.

As we mentioned earlier, the nontorsion case remains open in general. However, if FG is semiprime, then there is nothing more to do, as the reader can verify that the proof of the theorem below follows that of Theorem 4.6.1 virtually verbatim.

Theorem 5.3.4. *Let F be an infinite field of characteristic $p \neq 2$, and G a group containing an element of infinite order, such that FG is semiprime. If $\mathscr{U}^+(FG)$ is bounded Engel, then the torsion elements of G form a (normal) subgroup T such that*

1. *if $p = 0$, then T is abelian or a Hamiltonian 2-group;*
2. *if $p > 2$, then T is an abelian p'-group;*
3. *every idempotent of FT is central in FG; and*
4. *if H is a finite subgroup of G such that FH has a nonsymmetric primitive idempotent e, then G/T is bounded Engel and H/N is central in G/N, where N is the kernel of $g \mapsto ge$.*

Conversely, if G/T is a u.p. group and FG satisfies the four conditions above, then $\langle \mathscr{U}^+(FG) \rangle$ is bounded Engel.

Chapter 6
Solvability of $\mathscr{U}(FG)$ and $\mathscr{U}^+(FG)$

6.1 Introduction

Let us now consider the solvability of $\mathscr{U}(FG)$. The results for group rings of finite groups are classical, due to Bateman [4], Motose and Tominaga [80] and Motose and Ninomiya [79]. We will make the simplifying assumption that F has more than three elements. Essentially, the need for this comes down to the following well-known result due to Dickson and Jordan.

Proposition 6.1.1. *Let F be a field. Then $GL_2(F)$ is solvable if and only if $F = \mathbb{Z}_2$ or $\mathbb{Z}_3$.*

Proof. If $GL_2(F)$ is solvable, then so is its subhomomorphic image, $PSL_2(F)$. But if F has more than three elements, then certainly $PSL_2(F)$ is a nonabelian simple group (see, for instance, [89, 3.2.9]). On the other hand, if $F = \mathbb{Z}_2$ or $\mathbb{Z}_3$, then it is trivial to verify that $GL_2(F)$ is solvable. $\qquad\square$

There are several special cases to consider when F has 2 or 3 elements. More explicit descriptions of these cases were given by Bovdi and Khripta [12] and Passman [83]. We refer the reader to [83] or [54, Section 3.8] for an exposition of these results for finite groups.

Bovdi and Khripta also extended these methods to torsion groups in [13]. Again, we will present their results assuming that $|F| > 3$ and refer the reader to [10] for the $\mathbb{Z}_2$ and $\mathbb{Z}_3$ cases.

For nontorsion groups, Sehgal provided the answer for rational group algebras in [94]. For other fields, Bovdi proved some additional cases in [9] and then, making use of the results discussed in Chapter 1, proved the general case in [10]. (Of course, the usual restriction that G modulo its torsion part be a u.p. group must be imposed for the sufficiency.)

Naturally, we can also discuss the solvability of the symmetric units of FG. As $\mathscr{U}^+(FG)$ is not generally a group, we had better say what we mean. In any group, define

$$(g_1, g_2)^o = (g_1, g_2)$$

G.T. Lee, *Group Identities on Units and Symmetric Units of Group Rings*,
Algebra and Applications 12, DOI 10.1007/978-1-84996-504-0_6,
© Springer-Verlag London Limited 2010

and

$$(g_1,\ldots,g_{2^{n+1}})^o = ((g_1,\ldots,g_{2^n})^o,(g_{2^n+1},\ldots,g_{2^{n+1}})^o).$$

Then we say that $\mathscr{U}^+(FG)$ is solvable if there exists an n so that $(\alpha_1,\ldots,\alpha_{2^n})^o = 1$ for all $\alpha_i \in \mathscr{U}^+(FG)$.

In order to make use of the results in Chapter 2, we will assume that F is infinite and char $F = p \neq 2$. In [73], Lee and Spinelli established the conditions under which $\mathscr{U}^+(FG)$ is solvable if FG is semiprime or G has only finitely many p-elements. When G has infinitely many p-elements, it is possible to reduce our problem to the Lie solvability problem presented in Chapter 3; thus, a suitable restriction upon the orders of the group elements is necessary in order to make use of these results.

We will present the results on the solvability of $\mathscr{U}(FG)$ in the next section and on the solvability of $\mathscr{U}^+(FG)$ in the final section.

6.2 Solvable Unit Groups

Let us discuss the solvability of $\mathscr{U}(FG)$ when G is finite. The following observation is useful.

Lemma 6.2.1. *Let F be a field of characteristic $p \geq 2$ and G a group. Let N be a finite normal p-subgroup of G. Then $\mathscr{U}(FG)$ is solvable if and only if $\mathscr{U}(F(G/N))$ is solvable.*

Proof. The necessity follows from Lemma 1.2.18. Suppose that $\mathscr{U}(F(G/N))$ satisfies $(x_1,\ldots,x_{2^n})^o = 1$. Then letting $F\bar{G} = F(G/N)$, and taking any $\alpha_i \in \mathscr{U}(FG)$, we have $(\bar{\alpha}_1,\ldots,\bar{\alpha}_{2^n})^o = 1$. Thus, $(\alpha_1,\ldots,\alpha_{2^n})^o \in 1 + \Delta(G,N)$. Repeated applications of Lemma 4.7.3 reveal that, for any positive integer m,

$$(\alpha_1,\ldots,\alpha_{2^{n+m}})^o \in 1 + (\Delta(G,N))^{2^m}.$$

By Lemma 1.1.1, $\Delta(G,N)$ is nilpotent, and we conclude that $\mathscr{U}(FG)$ is solvable. □

The following theorem is from Bateman [4].

Theorem 6.2.2. *Let F be a field of characteristic $p \geq 0$, with $|F| > 3$. Let G be a finite group. Then $\mathscr{U}(FG)$ is solvable if and only if G is p-abelian.*

Proof. The characteristic zero case follows from Corollary 1.2.21. Thus, let p be prime. In view of the preceding lemma, we can factor out the maximal normal p-subgroup of G. Thus, we assume that G has no nontrivial normal p-subgroups and prove that if $\mathscr{U}(FG)$ is solvable, then G is abelian.

By Proposition 1.3.3, $J(FG)$ is nilpotent. Thus, we see from Lemma 1.2.17 that $\mathscr{U}(FG/J(FG))$ is solvable. But Proposition 1.3.3 tells us that $FG/J(FG) \cong \bigoplus_i M_{n_i}(D_i)$, where each $n_i \geq 1$ and each D_i is a division algebra over F. Thus, each $GL_{n_i}(D_i)$ is solvable. Therefore, by Proposition 1.2.2, each D_i is a field, and by Proposition 6.1.1 each $n_i = 1$. That is, $FG/J(FG)$ is commutative. Consider the group $H = G \cap (1 + J(FG))$. It is evidently a normal subgroup of G. Furthermore, if $h \in H$, then $h - 1$ is nilpotent; hence, h is a p-element. By our assumption on G, $H = 1$. Thus, the map $g \mapsto g + J(FG)$ embeds G in the abelian group $\mathscr{U}(FG/J(FG))$. We are done. $\square$

As an immediate consequence, we get

Corollary 6.2.3. *Let G be a torsion group and F a field of characteristic $p > 0$ such that $|F| > 3$. If $\mathscr{U}(FG)$ is solvable, then the p-elements of G form a (normal) subgroup P, and G/P is abelian.*

Proof. By Theorem 1.2.27, FG satisfies a polynomial identity. Thus, by Proposition 1.1.4, G is locally finite, and we can apply the preceding theorem to finite subgroups of G to obtain our result. $\square$

Let us consider torsion groups. If either char $F = 0$ or char $F = p > 0$ and G has only finitely many p-elements, then we are basically done. If G has infinitely many p-elements, then we will reduce our problem to the Lie solvable case. The following lemma is similar to Lemma 4.2.3, as is its proof.

Lemma 6.2.4. *Let R be any ring and let $\eta_1, \ldots, \eta_{2^n} \in R$ be central and square-zero. Then for any $r_1, \ldots, r_{2^n} \in R$, we have*

$$(1 + \eta_1 r_1, \ldots, 1 + \eta_{2^n} r_{2^n})^o = 1 + \eta_1 \cdots \eta_{2^n} [r_1, \ldots, r_{2^n}]^o.$$

Lemma 6.2.5. *Let F be a field of characteristic $p > 0$ containing more than three elements. Let G be a torsion group having infinitely many p-elements. If $\mathscr{U}(FG)$ is solvable, then FG is Lie solvable.*

Proof. By Theorem 1.2.27 and Proposition 1.1.4, G has a p-abelian normal subgroup N of finite index. In addition, by Lemma 1.2.18, $\mathscr{U}(F(G/N'))$ is solvable. Furthermore, in view of Theorem 3.1.3, FG is Lie solvable if and only if $F(G/N')$ is Lie solvable. Thus, we factor out N' and assume that N is abelian. Also, by Corollary 6.2.3 and Theorem 1.3.1, G' is a p-group of bounded exponent. If G' is finite, then we are done, by Theorem 3.1.3. Otherwise, Lemma 3.4.16 tells us that G contains an infinite direct product of nontrivial finite normal p-subgroups, $A = A_1 \times A_2 \times \cdots$.

Suppose that $\mathscr{U}(FG)$ satisfies $(x_1, \ldots, x_{2^n})^o = 1$. We claim that FG satisfies $[x_1, \ldots, x_{2^n}]^o = 0$. Suppose this is not the case, and choose $\alpha_1, \ldots, \alpha_{2^n} \in FG$ such that $[\alpha_1, \ldots, \alpha_{2^n}]^o \neq 0$. Let X be a transversal to A in G. Write $[\alpha_1, \ldots, \alpha_{2^n}]^o = \sum_j \beta_j g_j$, with each $\beta_j \in FA$, $g_j \in X$. We may choose an r so that all β_j lie in $F(A_1 \times A_2 \times \cdots \times A_r)$. For each $i \geq 1$, let $\eta_i = \hat{A}_{r+i}$. Each η_i is clearly central and square-zero. Furthermore, by the preceding lemma,

$$1 = (1 + \eta_1 \alpha_1, \ldots, 1 + \eta_{2^n} \alpha_{2^n})^o = 1 + \eta_1 \cdots \eta_{2^n} [\alpha_1, \ldots, \alpha_{2^n}]^o.$$

Thus,

$$\eta_1 \cdots \eta_{2^n} \sum_j \beta_j g_j = 0.$$

Now, the g_j lie in distinct cosets modulo A, so we must have

$$\eta_1 \cdots \eta_{2^n} \beta_j = 0,$$

for each j. But given that the product $A_1 \times A_2 \times \cdots$ is direct, and that each $\eta_i \neq 0$, we have $\beta_j = 0$ for all j. This contradicts the assumption that $[\alpha_1, \ldots, \alpha_{2^n}]^o \neq 0$. $\square$

There is still some work to do for the characteristic 2 case. To this end, let char $F = 2$ and suppose that G has an abelian subgroup A of index 2. Let $G = \langle A, g \rangle$, and let H be any torsion subgroup of $\mathscr{U}(FG)$. For each $m \geq 0$, we define

$$H_m = \{\alpha_1 + \alpha_2 g \in H : \alpha_i \in FA \text{ and } \alpha_i^{2^m} \text{ are central in } FG\}.$$

We claim that each H_m is a subgroup of H. Indeed, if $\alpha_1 + \alpha_2 g, \beta_1 + \beta_2 g \in H_m$, then their product is

$$(\alpha_1 \beta_1 + \alpha_2 \beta_2^{g^{-1}} g^2) + (\alpha_1 \beta_2 + \alpha_2 \beta_1^{g^{-1}}) g.$$

Evidently,

$$(\alpha_1 \beta_1 + \alpha_2 \beta_2^{g^{-1}} g^2)^{2^m} = \alpha_1^{2^m} \beta_1^{2^m} + \alpha_2^{2^m} (\beta_2^{2^m})^{g^{-1}} g^{2^{m+1}},$$

and since each of $\alpha_i^{2^m}$, $\beta_i^{2^m}$ and g^2 is central, this expression is central, and similarly for $(\alpha_1 \beta_2 + \alpha_2 \beta_1^{g^{-1}})^{2^m}$. As H is torsion, the claim is proved. Also, we have the following observation.

Lemma 6.2.6. *With H as above, for any positive integer m, we have*

1. *$\alpha^4, (\alpha^2 \beta^2)^2 \in H_{m-1}$ for all $\alpha, \beta \in H_m$; and*
2. *$H_m'' \leq H_{m-1}$.*

Proof. For the first part, let us consider α^4. Let $\alpha = \alpha_1 + \alpha_2 g$, with $\alpha_i \in FA$. Then, since char $F = 2$, we get

$$\alpha^4 = (\alpha_1^4 + (\alpha_1 + \alpha_1^{g^{-1}})^2 \alpha_2 \alpha_2^{g^{-1}} g^2 + (\alpha_2 \alpha_2^{g^{-1}})^2 g^4) + ((\alpha_1 + \alpha_1^{g^{-1}})^3 \alpha_2) g.$$

As $\alpha_1 + \alpha_1^{g^{-1}}$, $\alpha_2 \alpha_2^{g^{-1}}$ and $(\alpha_1^4)^{2^{m-1}}$ are all central, it remains only to check that $((\alpha_1 + \alpha_1^{g^{-1}})^3 \alpha_2)^{2^{m-1}}$ is central. But as $\alpha_1^{2^m}$ is central, we get $\alpha_1^{2^m} = (\alpha_1^{g^{-1}})^{2^m}$, hence $(\alpha_1 + \alpha_1^{g^{-1}})^{2^m} = 0$, and we are done.

The calculation for $(\alpha^2 \beta^2)^2$ is similar, but more unwieldy, and we suppress it.

To prove the second part, we make note of three identities that hold in any group; namely,

$$(a, b) = a^{-2} (b^{a^{-1}})^{-2} (ab)^2,$$

$$(ab,c) = (a,c)^b(b,c)$$

and

$$(a^2,b^2) = a^{-4}(b^{-4})^{a^{-2}}(a^2b^2)^2.$$

(The verification is routine.) By the first of these identities, every element of H''_m can be written as a product of squares of elements of H_m. Taking $\gamma, \delta \in H'_m$, let us say that $\gamma = \gamma_1^2 \cdots \gamma_r^2$, $\delta = \delta_1^2 \cdots \delta_s^2$, with each γ_i and $\delta_i \in H_m$. By the second identity, (γ, δ) is a product of terms of the form

$$(\gamma_i^2, \delta_j^2)^{\rho_k} = ((\gamma_i^{\rho_k})^2, (\delta_j^{\rho_k})^2)$$

for various $\rho_k \in H_m$. Letting $\tau = \gamma_i^{\rho_k}$, $\upsilon = \delta_j^{\rho_k}$, the third identity tells us that (γ, δ) is a product of terms of the form

$$\tau^{-4}(\upsilon^{\tau^{-2}})^{-4}(\tau^2\upsilon^2)^2.$$

Since $\tau, \upsilon \in H_m$, the first part of the lemma tells us that $(\gamma, \delta) \in H_{m-1}$. Thus, $H''_m \le H_{m-1}$, as required. □

Lemma 6.2.7. *Let F be a field of characteristic 2 and G a group having an abelian subgroup A of index 2. Let N be a normal 2-subgroup of G of bounded exponent contained in A. Then $1 + \Delta(G,N)$ is a solvable group.*

Proof. Let $H = 1 + \Delta(G,N)$. If $\alpha \in H$, then $\alpha - 1 = \sum_j \beta_j(n_j - 1)$, with each $\beta_j \in FG$, $n_j \in N$. As $A \le \phi(G)$, there exists a finite normal subgroup N_1 of G contained in N such that $n_j \in N_1$ for all j. Thus, $\alpha - 1 \in \Delta(G,N_1)$, and by Lemma 1.1.1, $\Delta(G,N_1)$ is nilpotent. Therefore, $\alpha^{2^r} = 1$ for some r, and H is a torsion group.

Suppose that N has exponent 2^m. Then clearly $\Delta(A,N)$ is nil of exponent 2^m. Writing $G = \langle A, g \rangle$ as above, take any $\alpha = \alpha_1 + \alpha_2 g \in H$. Then $\alpha_1 - 1, \alpha_2 \in \Delta(G,N) \cap FA = \Delta(A,N)$, hence $\alpha_1^{2^m} = 1$ and $\alpha_2^{2^m} = 0$. In particular, $\alpha \in H_m$, and therefore $H = H_m$. But using the second part of the previous lemma, and noting that H_0 is abelian, we see that H_m is solvable. We are done. □

We now present the proof of the torsion case, due to Bovdi and Khripta [13] and Bovdi [10].

Theorem 6.2.8. *Let F be a field of characteristic $p \ge 0$ and G a torsion group. Suppose that $|F| > 3$. Then $\mathcal{U}(FG)$ is solvable if and only if either*

1. *$p \ne 2$ and G is p-abelian; or*
2. *$p = 2$, G has a 2-abelian subgroup A of index at most 2, and G' is a 2-group of bounded exponent.*

Proof. If $p = 0$, then the result follows from Corollary 1.2.21, so assume that $p > 0$. Suppose that G has only finitely many p-elements. By Corollary 6.2.3, G is p-abelian. Conversely, by Lemma 6.2.1, since G' is a finite p-group, and $\mathcal{U}(F(G/G'))$ is abelian, $\mathcal{U}(FG)$ is solvable.

Therefore, we may assume that G has infinitely many p-elements. By Lemma 6.2.5, FG is Lie solvable. If $p > 2$, then Theorem 3.1.3 says that G is p-abelian. The converse follows as before. Therefore, let $p = 2$. By Theorem 3.1.3, G has a 2-abelian subgroup A of index at most 2. (If G is 2-abelian, then again, we are done, so assume that A has index 2.) By Corollary 6.2.3, G' is a 2-group. Theorem 1.3.1 completes the necessity.

Let us consider the sufficiency. By Lemma 6.2.1, it suffices to factor out A'. Thus, let A be abelian. If $G = A$, there is nothing to do, so let A have index 2. Now $G' \leq A$, hence, by Lemma 6.2.7, $1 + \Delta(G, G')$ is solvable. But the natural projection $FG \to F(G/G')$ induces a map $\mathscr{U}(FG) \to \mathscr{U}(F(G/G'))$ with kernel $1 + \Delta(G, G')$. As $\mathscr{U}(F(G/G'))$ is abelian, it follows that $\mathscr{U}(FG)$ is solvable. The proof is complete.   □

Combining Theorems 6.2.8 and 3.1.3, we immediately obtain

Corollary 6.2.9. *Let F be a field with char $F \neq 2, 3$ and let G be a torsion group. Then $\mathscr{U}(FG)$ is solvable if and only if FG is Lie solvable.*

The rest of this section deals with groups containing elements of infinite order. These results were proved by Bovdi in [9] and [10]. We begin with the semiprime case.

Theorem 6.2.10. *Let F be a field of characteristic $p \geq 0$, and G a group containing an element of infinite order. Let FG be semiprime. If $\mathscr{U}(FG)$ is solvable then*

1. the torsion elements form an abelian (normal) subgroup T of G;
2. if $p > 0$, then G is a p'-group;
3. every idempotent in FG is central; and
4. G is solvable.

Conversely, if G satisfies the four conditions above and G/T is a u.p. group, then $\mathscr{U}(FG)$ is solvable.

Proof. The necessity follows from Theorem 1.4.9 and the fact that if $\mathscr{U}(FG)$ is solvable, then so is G. Therefore, we need only check the sufficiency.

Suppose that G satisfies $(x_1, \ldots, x_{2^n})^o = 1$. We claim that $\mathscr{U}(FG)$ satisfies $(x_1, \ldots, x_{2^{n+1}})^o = 1$. Suppose this is not the case. Choose $\alpha_1, \ldots, \alpha_{2^{n+1}} \in \mathscr{U}(FG)$ such that $(\alpha_1, \ldots, \alpha_{2^{n+1}})^o \neq 1$. By Remark 1.4.10, there is a finite subgroup E of T such that for any primitive idempotent e of FE, we have $\alpha_i e = \lambda_i g_i$, where $\lambda_i \in \mathscr{U}(FEe)$ and $g_i \in G$. Now, $FE = FEe \oplus \cdots$ and, since e is central in FG, $FG = FGe \oplus \cdots$. Thus, it suffices to show that

$$(\alpha_1 e, \ldots, \alpha_{2^{n+1}} e)^o = e$$

for all such e. It is clear that $\mathscr{U}(FEe)$ is a normal subgroup of $\langle \mathscr{U}(FEe), Ge \rangle$. Indeed, this follows from the fact that E is a normal subgroup of G (since $\frac{1}{|E|}\hat{E}$ is an idempotent, hence central). Thus,

$$(\alpha_1 e, \ldots, \alpha_{2^n} e)^o = (\lambda_1 g_1, \ldots, \lambda_{2^n} g_{2^n})^o \equiv (g_1, \ldots, g_{2^n})^o e \pmod{\mathscr{U}(FEe)}.$$

But $(g_1, \ldots, g_{2^n})^o = 1$, hence $(\alpha_1 e, \ldots, \alpha_{2^n} e)^o \in \mathscr{U}(FEe)$, which is abelian. As the same is true for $(\alpha_{2^n+1}, \ldots, \alpha_{2^{n+1}})^o$, the claim is proved, and we are done.   □

We can now focus on the prime characteristic case.

Lemma 6.2.11. *Let char $F = p > 0$ and let G be a nontorsion group such that $\mathscr{U}(FG)$ is solvable. Then the p-elements form a (normal) subgroup P of G. Furthermore, if P is infinite, then FG is Lie solvable.*

Proof. The fact that P is a subgroup follows comes from Propositions 1.5.2 and 1.5.5. Let P be infinite. Then, by Theorems 1.5.10 and 1.5.16, G has a p-abelian subgroup of finite index. If P has bounded exponent, then we follow the proof of Lemma 6.2.5 verbatim. Otherwise, by Theorem 1.5.16, G' is a p-group of bounded exponent. If G' is a finite p-group, then we are done, by Theorem 3.1.3. Otherwise, we again have an infinite p-subgroup of bounded exponent, and the proof of Lemma 6.2.5 does the job.   □

If P is finite, there is nothing more to do. Lemma 6.2.1 says that $\mathscr{U}(FG)$ is solvable if and only if $\mathscr{U}(F(G/P))$ is solvable, and by Proposition 1.2.9, $F(G/P)$ is semiprime. Thus, Theorem 6.2.10 applies.

Theorem 6.2.12. *Let F be a field of characteristic $p > 0$ and G a nontorsion group containing finitely many p-elements. Then $\mathscr{U}(FG)$ is solvable if and only if the p-elements of G form a (normal) subgroup P and $\mathscr{U}(F(G/P))$ is solvable.*

If P is infinite, then the odd prime characteristic case is done as well.

Theorem 6.2.13. *Let F be a field of characteristic $p > 2$ and G a nontorsion group containing infinitely many p-elements. Then $\mathscr{U}(FG)$ is solvable if and only if G is p-abelian.*

Proof. Suppose that $\mathscr{U}(FG)$ is solvable. By Lemma 6.2.11, FG is Lie solvable, and Theorem 3.1.3 completes the necessity. The sufficiency follows immediately from Lemma 6.2.1.   □

If $p = 2$, there is still some work to do, as G is not necessarily 2-abelian. We must break the result down into two cases.

Theorem 6.2.14. *Let F be a field of characteristic 2, and let G be a nontorsion group containing infinitely many 2-elements. Suppose that the 2-elements are of bounded exponent. Then $\mathscr{U}(FG)$ is solvable if and only if*

1. *G has a 2-abelian subgroup A of index at most 2;*
2. *the 2-elements of G form a (normal) subgroup P;*
3. *the torsion elements of G/P form an abelian group, T/P; and*
4. *every idempotent of $F(G/P)$ is central.*

Proof. Suppose that $\mathcal{U}(FG)$ is solvable. Then the existence of A follows from Lemma 6.2.11 and Theorem 3.1.3, and the remaining conditions come from Theorem 1.5.10.

Let us consider the converse. By Lemma 6.2.1, it is enough to show that $\mathcal{U}(F(G/A'))$ is solvable; hence, we factor out A' and assume that A is abelian. If $G = A$, then there is nothing to do, so assume that $(G : A) = 2$. By Lemma 6.2.7, $1 + \Delta(G, A \cap P)$ is solvable. But we have the obvious homomorphism $\mathcal{U}(FG) \to \mathcal{U}(F(G/(A \cap P)))$ with kernel $1 + \Delta(G, A \cap P)$. Thus, it suffices to show that $\mathcal{U}(F(G/(A \cap P)))$ is solvable. But $(P : A \cap P) \le (G : A) = 2$. Thus, $G/(A \cap P)$ has at most two 2-elements. By Lemma 6.2.1, it suffices to factor out the 2-subgroup of $G/(A \cap P)$ and assume that G is a $2'$-group. In order to apply Theorem 6.2.10, it remains to check that G is solvable and G/T is a u.p. group. But the existence of an abelian subgroup of index 2 establishes that G is solvable. Furthermore, by [82, Lemma 13.3.1], every finitely generated subgroup of G/T is right orderable, hence a u.p. group. Thus, G/T is a u.p. group. We are done.   □

The last of the main results of [10] is

Theorem 6.2.15. *Let F be a field of characteristic 2 and G a nontorsion group whose 2-elements have unbounded exponent. Then $\mathcal{U}(FG)$ is solvable if and only if G has a 2-abelian subgroup A of index at most 2, and G' is a 2-group of bounded exponent.*

Proof. Let $\mathcal{U}(FG)$ be solvable. By Lemma 6.2.11, FG is Lie solvable, hence, by Theorem 3.1.3, G has a 2-abelian subgroup of index at most 2. Theorem 1.5.16 completes the proof of the necessity.

For the sufficiency, we note that by Lemma 6.2.1, it suffices to factor out A' and assume that A is abelian. If $G = A$, there is nothing to do, so let A have index 2. Now $G' \le A$ hence, by Lemma 6.2.7, $1 + \Delta(G, G')$ is solvable. Since we have the natural map $\mathcal{U}(FG) \to \mathcal{U}(F(G/G'))$ with kernel $1 + \Delta(G, G')$, and $\mathcal{U}(F(G/G'))$ is abelian, it follows that $\mathcal{U}(FG)$ is solvable. We are done.   □

6.3 Solvable Symmetric Units

We now consider groups G such that $\mathcal{U}^+(FG)$ satisfies a solvability identity, $(x_1, \ldots, x_{2^n})^o = 1$. These results are taken from Lee and Spinelli [73]. In order to make use of the classifications in Chapter 2, we assume that F is infinite and char $F = p \ne 2$.

We begin with the following lemma, which is the analogue of Lemma 6.2.1 for symmetric units.

Lemma 6.3.1. *Let char $F = p > 2$ and let G have a finite normal p-subgroup N. Then $\mathcal{U}^+(FG)$ is solvable if and only if $\mathcal{U}^+(F(G/N))$ is solvable.*

Proof. For the necessity, see Lemma 2.3.6. For the sufficiency, follow the proof of Lemma 6.2.1, simply making all of the units symmetric.   □

This allows us to take care of the $p = 0$ case, as well as the case with finitely many p-elements, for torsion groups.

Theorem 6.3.2. *Let F be an infinite field of characteristic $p \neq 2$ and G a torsion group. If $p > 2$, let G have finitely many p-elements. Then $\mathscr{U}^+(FG)$ is solvable if and only if either*

1. $p = 0$ *and G is abelian or a Hamiltonian 2-group; or*
2. $p > 2$, *the p-elements of G form a normal subgroup P, and G/P is abelian or a Hamiltonian 2-group.*

Proof. The necessity follows from Theorems 2.4.8 and 2.4.9. For the sufficiency, combine the preceding lemma with Lemma 2.1.1. □

If G is not torsion, then the solution to the semiprime case is the following.

Theorem 6.3.3. *Let F be an infinite field of characteristic $p \neq 2$ and let G be a group containing an element of infinite order. Suppose that FG is semiprime. If $\mathscr{U}^+(FG)$ is solvable, then*

1. *if $p = 0$, then the set of torsion elements, T, is a (normal) subgroup of G, and T is abelian or a Hamiltonian 2-group;*
2. *if $p > 2$, then T is an abelian (normal) p'-subgroup of G;*
3. *every idempotent in FT is central in FG; and*
4. *if FT contains a nonsymmetric idempotent, then G is solvable.*

Conversely, if G/T is a u.p. group and FG satisfies the above four conditions, then $\mathscr{U}^+(FG)$ is solvable.

Proof. Suppose that $\mathscr{U}^+(FG)$ is solvable. Then the first three conditions follow from Theorem 2.5.6, so let us establish the fourth. Suppose that FT has a nonsymmetric idempotent. Let H be the (finite) subgroup of T generated by the support of this idempotent. Then FH has a nonsymmetric primitive idempotent, e. Now, e^* is also a primitive idempotent of FH, hence

$$FH = FHe \oplus FHe^* \oplus \cdots$$

and, since e is central in FG,

$$FG = FGe \oplus FGe^* \oplus \cdots.$$

Thus, for any $g \in G$, $ge + g^{-1}e^* + (1 - (e + e^*)) \in \mathscr{U}^+(FG)$. If $\mathscr{U}^+(FG)$ satisfies $(x_1, \ldots, x_{2^n})^o = 1$, then looking at the FGe component, we obtain

$$(g_1e, \ldots, g_{2^n}e)^o = e$$

for all $g_i \in G$. Thus, $(g_1, \ldots, g_{2^n})^o$ lies in the kernel of $g \mapsto ge$, which is contained in T. In particular, G/T is solvable. As T is abelian or a Hamiltonian 2-group, it is nilpotent. Thus, G is solvable.

Let us now consider the converse. Theorem 6.2.10 finishes the proof unless either T is a Hamiltonian 2-group or every idempotent in FT is symmetric. As every idempotent in FT is central, Lemma 2.1.1 tells us that if T is a Hamiltonian 2-group, then every idempotent in FT is symmetric anyway. But in this case, we apply Remark 2.5.7 in order to see that the symmetric units in FG commute.   □

The characteristic zero case is now complete, so assume that char $F = p > 2$ and that G has p-elements. Suppose that $\mathscr{U}^+(FG)$ is solvable. If there are only finitely many p-elements, then by Proposition 2.6.3, they form a subgroup P of G. In view of Lemma 6.3.1, we can reduce to the semiprime case, $F(G/P)$. We have proved

Theorem 6.3.4. *Let F be an infinite field with char $F = p > 2$. Let G be a nontorsion group having finitely many p-elements. Then $\mathscr{U}^+(FG)$ is solvable if and only if the p-elements of G form a (normal) subgroup P and $\mathscr{U}^+(F(G/P))$ is solvable.*

Finally, let us consider groups (torsion or nontorsion) having infinitely many p-elements. In fact, all of the work here was presented in Chapter 3, because we can reduce FG to a group ring whose symmetric elements are Lie solvable.

Lemma 6.3.5. *Let F be an infinite field with char $F = p > 2$. Let G be a group with infinitely many p-elements. If $\mathscr{U}^+(FG)$ is solvable, then $(FG)^+$ is Lie solvable, and either G is p-abelian or G has an infinite p-subgroup H of bounded exponent.*

Proof. Suppose that the p-elements of G have bounded exponent. Then by Proposition 2.4.3 and Theorem 2.6.5, they form a subgroup, P (hence we can use $H = P$), and G has a p-abelian normal subgroup A of finite index. By Lemma 6.3.1, $\mathscr{U}^+(F(G/A'))$ is solvable. Furthermore, by Lemma 3.4.8, $(FG)^+$ is Lie solvable if and only if $(F(G/A'))^+$ is Lie solvable. Thus, we factor out A' and assume that A is abelian. Since P has bounded exponent, by Lemma 3.4.16 G contains an infinite direct product of nontrivial finite abelian normal p-subgroups.

If P has unbounded exponent, then Theorems 2.4.8, 2.4.9 and 2.6.11 again give us A, and say that G' is a p-group of bounded exponent. If G' is finite, then we are done, by Theorem 3.1.3. Otherwise, we let $H = G'$ and again apply Lemma 3.4.16 to obtain an infinite direct product of nontrivial finite abelian normal p-subgroups.

For the rest, simply follow the proof of Lemma 6.2.5, taking $\alpha_i \in (FG)^+$ and noting that each η_i is symmetric.   □

Thus, we obtain the result on groups containing infinitely many p-elements from [73].

Theorem 6.3.6. *Let F be an infinite field of characteristic $p > 2$ and G a group containing infinitely many p-elements but no nontrivial elements of order dividing $p^2 - 1$. Then the following are equivalent:*

(i) $\mathscr{U}^+(FG)$ is solvable;
(ii) $\mathscr{U}(FG)$ is solvable;
(iii) G is p-abelian.

Proof. In view of Theorems 6.2.8 and 6.2.13, it remains only to check that (i) implies (iii). Suppose that $\mathscr{U}^+(FG)$ is solvable. Then apply the preceding lemma. If G is not p-abelian, then use Theorem 3.4.15 to obtain a contradiction. □

We conjecture that this theorem is true under the weaker assumption that G has no 2-elements, but in general, this problem remains open for torsion groups. Of course, using Theorem 3.4.1 in place of Theorem 3.4.15 in the proof of the last theorem, we obtain

Theorem 6.3.7. *Let F be an infinite field of characteristic $p > 2$ and G a nontorsion group containing infinitely many p-elements but no 2-elements. Then the following are equivalent:*

(i) $\mathscr{U}^+(FG)$ is solvable;
(ii) $\mathscr{U}(FG)$ is solvable;
(iii) G is p-abelian.

Chapter 7
Further Reading

7.1 Introduction

In this short chapter we mention a few results not discussed elsewhere in the book. No proofs are given.

The next section contains theorems concerning identities related to those studied earlier. For example, we discuss when $\mathscr{U}(FG)$ satisfies $(x_1,\ldots,x_n)=1$ for a particular n and when $\mathscr{U}(FG)$ is n-Engel for a particular n. We also discuss some generalizations of these concepts. In addition, we present another identity due to Cliff and Sehgal [24] and Coelho [25], and one due to Sahai [93].

In the final section, we provide references to some interesting new results concerning elements symmetric with respect to an involution $*$ other than the classical one.

7.2 Other Identities

Here we refer the reader to some results on identities that have not been proved in this book. To the best of the author's knowledge, with one exception noted below, no analogous theorems have been established for the symmetric units. To be sure, any such results would be welcome.

We begin by considering special cases of the nilpotency of $\mathscr{U}(FG)$. Of course, by the nilpotency class of $\mathscr{U}(FG)$ we mean the smallest positive integer n such that $\mathscr{U}(FG)$ satisfies $(x_1,\ldots,x_{n+1})=1$. While the nilpotency class of $\mathscr{U}(FG)$ is not known for all G, there are several notable results.

Let us begin with the semiprime case. It is clear from the proof of Theorem 4.2.9 that if the torsion elements are central in G then the nilpotency class of $\mathscr{U}(FG)$ equals that of G. In fact, in [60, Theorem], Kurdics also dealt with the exceptional case where $|F|$ is a Mersenne prime. His result is

G.T. Lee, *Group Identities on Units and Symmetric Units of Group Rings*,
Algebra and Applications 12, DOI 10.1007/978-1-84996-504-0_7,
© Springer-Verlag London Limited 2010

Theorem 7.2.1. *Let F be a field of order $p = 2^q - 1$, a Mersenne prime, and let G be a nilpotent group of class n containing no p-elements. Suppose that G contains a noncentral element of finite order. If $\mathscr{U}(FG)$ is nilpotent, then its nilpotency class is the greater of n and $q + 1$.*

Thus, let us assume that char $F = p > 0$ and G has p-elements. Evidently, the $n = 1$ case is uninteresting. Indeed, $\mathscr{U}(FG)$ is abelian if and only if G is abelian. The $n = 2$ case is complete. Certainly, if $\mathscr{U}(FG)$ is nilpotent of class 2, then so is G. In [56] Khripta proved the following more general result.

Theorem 7.2.2. *Let F be a field of characteristic $p > 0$ and G a nilpotent group of class $n > 1$. Let P be the group of p-elements of G and suppose that $P \neq 1$. Then $\mathscr{U}(FG)$ is nilpotent of class n if and only if one of the following holds:*

1. $n = p = 2$ and $|G'| = 2$.
2. $n = p = 2$ and $G' = P \simeq C_2 \times C_2$.
3. $n = 2$, $p = 3$ and $|G'| = |P| = 3$.
4. $n = 3$, $p = 2$ and $|G'| = |P| = 4$.

Proof. See [56, Theorem 2].   □

What of the $n = 3$ case? Clearly, if $\mathscr{U}(FG)$ has nilpotency class 3, then G has nilpotency class 2 or 3. The latter case is covered by the fourth part of the preceding theorem, so it remains to consider groups of nilpotency class 2. When G is a finite p-group, $p > 2$, this problem was the subject of a series of papers by Shalev [98, 99, 101] and Mann and Shalev [78]. The finite 2-groups were then handled in Rao and Sandling [85]. This theorem was then extended to arbitrary groups in Bovdi and Kurdics [16]. The general result is as follows.

Theorem 7.2.3. *Let F be a field of characteristic $p > 0$ and G a nilpotent group. Let P be the group of p-elements of G, and suppose that $P \neq 1$. Then $\mathscr{U}(FG)$ is nilpotent of class 3 if and only if either $p = 3$, $|G'| = 3$ and $G' \neq P$, or $p = 2$ and one of the following holds:*

1. $G' = P \simeq C_4$.
2. G is nilpotent of class 2 and $G' = P \simeq C_2 \times C_2 \times C_2$.
3. G is nilpotent of class 2, $G' \simeq C_2 \times C_2$ and $G' \neq P$.
4. G is nilpotent of class 3 and $G' = P \simeq C_2 \times C_2$.
5. $G' \simeq C_2 \times C_2 \times C_2$, $|P| = 16$, P is central in G and no conjugacy class in G has more than 4 elements.

Proof. See [16, Theorem 5.2].   □

While there is no complete result for larger nilpotency classes, the papers cited above do contain numerous interesting partial results in this direction, and we refer the reader to them. Also of interest is the following result of Kurdics. (See Coleman and Passman [26] for an earlier result for finite p-groups.)

Theorem 7.2.4. *Let char $F = p > 2$, and let G be a nonabelian nilpotent group with G' a finite p-group. Then the nilpotency class of $\mathscr{U}(FG)$ is at least $p - 1$. Furthermore, if P is the set of p-elements of G, then*

1. *$\mathscr{U}(FG)$ is nilpotent of class $p - 1$ if and only if $|P| = p$; and*
2. *$\mathscr{U}(FG)$ is nilpotent of class p if and only if $|G'| = p$ and $G' \neq P$.*

Proof. See [61, Corollary 1]. $\square$

We mention one other useful reduction. Proposition 5.2.9 tells us that if FG satisfies $[x_1, \ldots, x_n] = 0$, then $\mathscr{U}(FG)$ satisfies $(x_1, \ldots, x_n) = 1$. But from Du [29] (for finite p-groups) and Catino et al. [23] (for arbitrary torsion groups), we obtain this more specific version of Corollary 4.2.7.

Theorem 7.2.5. *Let F be a field and G a torsion group. Fix an integer $n \geq 2$. Then $\mathscr{U}(FG)$ satisfies the group identity $(x_1, \ldots, x_n) = 1$ if and only if FG satisfies the polynomial identity $[x_1, \ldots, x_n] = 0$.*

Proof. See [23, Theorem]. $\square$

This theorem can be used in combination with the main result of Bhandari and Passi [7], who presented a formula for the Lie nilpotency index of FG when char $F > 3$.

Considering a somewhat weaker condition upon the unit group, we can ask when $\mathscr{U}(FG)$ is hypercentral. Of course, hypercentrality is not a group identity, but it is certainly a natural generalization. The main result of Riley [86] is the following.

Theorem 7.2.6. *Let F be a field of characteristic $p > 0$ and G a group having an element of order p. Then the following are equivalent:*

(i) *$\mathscr{U}(FG)$ is hypercentral;*
(ii) *$\mathscr{U}(FG)$ is nilpotent;*
(iii) *G is nilpotent and p-abelian.*

In [93], Sahai looked at the conditions under which $(\mathscr{U}(FG))'$ is nilpotent of class at most 2. In view of Lemma 4.1.2, this is equivalent to $\mathscr{U}(FG)$ satisfying the identity $((x_1, x_2), (x_3, x_4), (x_5, x_6)) = 1$. The main result is

Theorem 7.2.7. *Suppose char $F = p \neq 2$ and G is a finite nonabelian group. Then $(\mathscr{U}(FG))'$ is nilpotent of class at most 2 if and only if one of the following holds:*

1. *$p = 3$ and $|G'| = 3$.*
2. *$p = 3$, G is nilpotent of class 2 and $G' \simeq C_3 \times C_3$.*
3. *$p = 5$, G' is central and $|G'| = 5$.*

Proof. See [93, Theorem 4.2]. $\square$

Corollary 7.2.8. *Suppose char $F \neq 2$ and G is a finite group. Then $\mathscr{U}(FG)$ satisfies the group identity $((x_1, x_2), (x_3, x_4), (x_5, x_6)) = 1$ if and only if FG satisfies the polynomial identity $[[x_1, x_2], [x_3, x_4], [x_5, x_6]] = 0$.*

Proof. See [93, Corollary 4.3].   □

Kurdics [58, 60, 61] proved theorems similar to those above for the n-Engel property. Two such results are these:

Theorem 7.2.9. *Let char $F = p > 0$ and let G be a nonabelian nilpotent group. Let P be the group of p-elements of G, and suppose that $P \neq 1$. Then $\mathscr{U}(FG)$ is 2-Engel if and only if one of the following holds:*

1. *$p = 2$ and $|G'| = 2$.*
2. *$p = 2$, G is nilpotent of class 2 and $G' = P \simeq C_2 \times C_2$.*
3. *$p = 3$ and $|G'| = |P| = 3$.*

Proof. See [61, Theorem 3].   □

Theorem 7.2.10. *Let char $F = p > 2$ and let G be a nonabelian nilpotent group. Let P be the group of p-elements of G and suppose that $P \neq 1$. Then $\mathscr{U}(FG)$ is not $(p - 2)$-Engel. Furthermore,*

1. *$\mathscr{U}(FG)$ is $(p - 1)$-Engel if and only if $|G'| = |P| = p$, and*
2. *$\mathscr{U}(FG)$ is p-Engel if and only if $|G'| = p$.*

Proof. See [61, Theorems 1 and 2].   □

The result for 3-Engel groups has rather a lot of cases, and we refer the reader to [61, Theorem 4].

We can also look a little beyond the notion of a group identity here and consider Engel groups. Recall that a group G is said to be Engel if, for every $g, h \in G$, there exists an n (possibly depending upon g and h) such that

$$(g, \underbrace{h, \ldots, h}_{n \text{ times}}) = 1.$$

Suppose that G has no elements of order char F. Then Bovdi and Khripta [14, 15] determined the groups such that $\mathscr{U}(FG)$ is Engel, and the solution is the same as Theorem 5.2.1, *mutatis mutandis* (see [11, Theorem 1.1]). There is no complete solution for modular group algebras. A solution was, however, given for solvable groups in [14] and [15], and extended to the following form by Bovdi in [11].

Theorem 7.2.11. *Let F be a field of characteristic $p > 0$ and G a group having a nontrivial Sylow p-subgroup, P. Suppose either (i) G is solvable or (ii) P is solvable, normal in G, and P contains a nontrivial finite subgroup that is normal in G. Then $\mathscr{U}(FG)$ is Engel if and only if G is locally nilpotent and G' is a p-group.*

Proof. See [11, Theorem 3.2].   □

We can also ask about special cases of the solvability of $\mathscr{U}(FG)$. Recall that a group is said to be metabelian if it satisfies the group identity $(x_1, x_2, x_3, x_4)^o = 1$. Shalev [100] classified the finite groups G such that $\mathscr{U}(FG)$ is metabelian, when char $F \neq 2$. The characteristic 2 case was handled independently by Coleman and Sandling in [27] and Kurdics in [59]. The result is as follows.

Theorem 7.2.12. *Let F be a field of characteristic $p \geq 0$ and G a finite nonabelian group. Then $\mathscr{U}(FG)$ is metabelian if and only if one of the following holds:*

1. *$p = 3$, G is nilpotent of class 2 and $G' \simeq C_3$.*
2. *$p = 2$, G is nilpotent of class 2 and $G' \simeq C_2$ or $C_2 \times C_2$.*
3. *$F = \mathbb{Z}_2$ and $G = N \rtimes \langle g \rangle$, where N is an elementary abelian 3-group, g has order 2 and $g^{-1}ag = a^{-1}$ for all $a \in N$.*

Proof. See [27, Theorem 1.2].   □

For other results concerning the derived length of $\mathscr{U}(FG)$, see Bagiński [1] and Balogh and Li [3]. Also see Balogh and Juhász [2] for a related result on the symmetric units.

Let us also ask when $\mathscr{U}(FG)$ has bounded exponent modulo its centre; that is, when it satisfies the group identity $(x_1^n, x_2) = 1$ for some positive integer n. When char $F = 0$, Cliff and Sehgal [24] provided the answer for solvable groups G.

Theorem 7.2.13. *Let char $F = 0$ and let G be a solvable group. Then $\mathscr{U}(FG)$ has bounded exponent modulo its centre if and only if G has bounded exponent modulo its centre, and the torsion elements of G are central.*

Proof. See [24, Theorem 2].   □

Of course, if G is torsion, then in view of Corollary 1.2.21, the assumption that G is solvable can be dropped in the above theorem. If char $F = p > 0$, then the solution was found in [24] if G is solvable and n is either a p-power or relatively prime to p. Coelho [25] then extended this to all values of n and groups G that are locally finite, solvable or FC. As we now know (see Theorem 1.2.27 and Proposition 1.1.4) that if G is torsion and $\mathscr{U}(FG)$ satisfies a group identity, then G is locally finite, we do not need any restrictions when G is torsion.

Theorem 7.2.14. *Let char $F = p > 0$ and let G be a torsion group. Then $\mathscr{U}(FG)$ is of bounded exponent modulo its centre if and only if*

1. *G is of bounded exponent modulo its centre;*
2. *G has a p-abelian normal subgroup of finite index; and*
3. *either every p'-element of G is central, or G has bounded exponent and F is a finite field.*

Proof. See [25, Theorem A].   □

Let us now consider nontorsion groups. We let T be the set of torsion elements of G, and let P and Q be the sets of p-elements and p'-elements in T, respectively.

Theorem 7.2.15. *Let char $F = p > 0$ and let G be either solvable or an FC-group. Suppose that G has elements of infinite order. Then $\mathscr{U}(FG)$ has bounded exponent modulo its centre if and only if either G is nilpotent and has a p-abelian normal subgroup of finite p-power index, or*

1. G has bounded exponent modulo its centre;
2. P and Q are subgroups of G, with P of bounded exponent and Q central in T;
3. if Q is not central in G, then F is finite, Q has bounded exponent, and for every $g \in G$, $a \in Q$, there exists a positive integer r divisible by the degree of F over $\mathbb{Z}_p$, such that $a^g = a^{p^r}$; and
4. if P is infinite, then G has a p-abelian normal subgroup of finite index.

Proof. See [25, Theorem C]. $\qquad\qquad\qquad\qquad\qquad\qquad\qquad\qquad\qquad\qquad\qquad\qquad\square$

Cliff and Sehgal also extended this beyond the realm of group identities and considered when $\mathscr{U}(FG)/\zeta(\mathscr{U}(FG))$ is torsion. Adopting the same notation as above, their result is

Theorem 7.2.16. *Let char $F = p \geq 0$ and let G be solvable. Then $\mathscr{U}(FG)$ is torsion modulo its centre if and only if G is torsion modulo its centre and one of the following holds:*

1. $\mathscr{U}(FG)$ *is torsion.*
2. *If $p = 0$ (resp., $p > 0$), then T (resp., Q) is central in G.*
3. *$p > 0$, F is algebraic over $\mathbb{Z}_p$ and every idempotent in FG is central.*

Proof. See [24, Theorem 1]. $\qquad\qquad\qquad\qquad\qquad\qquad\qquad\qquad\qquad\qquad\qquad\qquad\square$

7.3 Other Involutions

Up until now, we have always worked with the involution given by $(\sum_{g \in G} \alpha_g g)^* = \sum_{g \in G} \alpha_g g^{-1}$. Indeed, until quite recently, little work had been done on elements symmetric with respect to other involutions on FG. During the time when this book was being written, however, several new results considering other involutions were proved. We mention some such results now.

Throughout this section, we write $(FG)^+$ *for the set of elements symmetric with respect to an involution $*$ on FG, and $\mathscr{U}^+(FG)$ for the set of units symmetric with respect to this involution. We do not assume that $*$ is the "classical" involution used in previous chapters.*

Take any field F and group G. Allowing $*$ to be a completely general involution on FG seems too ambitious, so let us consider involutions on FG induced from involutions on G. That is, let $* : G \to G$ be a function satisfying $(gh)^* = h^*g^*$ and $(g^*)^* = g$ for all $g, h \in G$. Extending this F-linearly, we obtain an involution on FG. (Evidently, the classical involution is the one induced from $g \mapsto g^{-1}$ on G.)

Working in this setting is more difficult than working with the classical involution. Indeed, if char $F \neq 2$, then we are accustomed to assuming that if N is a normal subgroup of G, then $(F(G/N))^+$ is a homomorphic image of $(FG)^+$. Thus, if $(FG)^+$ satisfies a particular polynomial identity, then so does $(F(G/N))^+$. But it may not even make sense to ask if this is true if $*$ is a different involution. Indeed, we need to be sure that G/N has an induced involution; that is, we must know that

N is $*$-invariant. Furthermore, if $(FG)^+$ is bounded Lie Engel and char $F = p > 0$, then one of our principal tools was the equation

$$0 = [g + g^{-1}, \underbrace{h + h^{-1}, \ldots, h + h^{-1}}_{p^n \text{ times}}] = [g + g^{-1}, h^{p^n} + h^{-p^n}].$$

If $*$ is not classical, then we still have

$$0 = [g + g^*, \underbrace{h + h^*, \ldots, h + h^*}_{p^n \text{ times}}] = [g + g^*, (h + h^*)^{p^n}],$$

but this is of limited use unless h and h^* commute.

The first major result for this sort of involution was given by Jespers and Ruiz Marín in [51]. They found the groups G such that $(FG)^+$ is commutative. In order to state the result, some definitions are in order. First, we recall that an LC-group (short for "lack of commutativity") is a nonabelian group G such that if $g, h \in G$ satisfy $gh = hg$, then at least one element of $\{g, h, gh\}$ lies in the centre of G. These groups were introduced by Goodaire. By Goodaire et al. [44, Proposition III.3.6], a group G is an LC-group with a unique nonidentity commutator if and only if $G/\zeta(G) \simeq C_2 \times C_2$. If G has an involution $*$, then we say that G is a special LC-group, or SLC-group, if it is an LC-group, it has a unique nonidentity commutator z, and for all $g \in G$ we have $g^* = g$ if $g \in \zeta(G)$ and, otherwise, $g^* = zg$. For example, if $*$ is the classical involution, then it is easy to see that the SLC-groups are precisely the Hamiltonian 2-groups. The result is the following.

Theorem 7.3.1. *Let F be a field of characteristic different from 2 and G a group with involution $*$. Let FG have the induced involution. Then the following are equivalent:*

(i) $(FG)^+$ is commutative;
(ii) $(FG)^+$ is central in FG;
(iii) G is abelian or an SLC-group.

Proof. See [51, Theorem 2.4]. □

We note that Jespers and Ruiz Marín also found the solution when char $F = 2$. For the classical involution, the classification is due to Broche Cristo [19].

The classifications from Chapter 3 of groups G such that $(FG)^+$ is Lie nilpotent or bounded Lie Engel have now been extended beyond the classical involution. This work was begun by Giambruno et al. in [33]. They proved

Lemma 7.3.2. *Let R be a semiprime ring with involution such that $2R = R$. If R^+ is bounded Lie Engel, then R^+ is commutative.*

Proof. See [33, Lemma 2.4]. □

In particular, if char $F = 0$, then by Proposition 1.2.9, FG is semiprime. Thus, Theorem 7.3.1 and Lemma 7.3.2 completely determine when $(FG)^+$ is bounded Lie Engel or Lie nilpotent. Let us consider the odd prime characteristic case.

Lemma 7.3.3. *Let FG have an involution induced from one on G. If char $F = p > 2$ and $(FG)^+$ is bounded Lie Engel, then the p-elements of G form a (normal) subgroup of G.*

Proof. See [33, Proposition 3.2]. □

Write P for the set of p-elements of G. Evidently P is $*$-invariant and therefore, if $(FG)^+$ is bounded Lie Engel (resp., Lie nilpotent), then so is $(F(G/P))^+$. By Proposition 1.2.9, $F(G/P)$ is semiprime. Thus, we may assume that G/P is abelian or an SLC-group. The following result was proved in Giambruno et al. [33] for groups without 2-elements, and in Lee et al. [70] for groups with 2-elements.

Theorem 7.3.4. *Let FG have an involution induced from one on G. Let char $F = p > 2$ and let G be a group such that G/P is abelian. If $(FG)^+$ is Lie nilpotent (resp., bounded Lie Engel), then FG is Lie nilpotent (resp., bounded Lie Engel).*

Proof. See [70, Propositions 1 and 2]. □

Thus, we may as well assume that G/P is an SLC-group. This case was handled in Lee et al. [70].

Theorem 7.3.5. *Let F be a field of characteristic $p > 2$ and G a group such that G/P is an SLC-group. Let FG have an involution induced from one on G. Then $(FG)^+$ is Lie nilpotent if and only if G is nilpotent and G has a finite normal $*$-invariant p-subgroup N such that G/N is an SLC-group.*

Proof. See [70, Theorem 1]. □

Theorem 7.3.6. *Let F be a field of characteristic $p > 2$ and G a group such that G/P is an SLC-group. Let FG have an involution induced from one on G. Then $(FG)^+$ is bounded Lie Engel if and only if G is nilpotent, G has a p-abelian $*$-invariant normal subgroup of finite index, and G has a $*$-invariant normal p-subgroup N of bounded exponent such that G/N is an SLC-group.*

Proof. See [70, Theorem 2]. □

Remark 7.3.7. The skew elements of FG under these involutions have also been considered. See Broche Cristo et al. [20] for the conditions under which $(FG)^-$ is commutative, and Giambruno et al. [35] for the conditions under which $(FG)^-$ is Lie nilpotent, with suitable restrictions upon G.

What of the set of symmetric units? It seems natural to ask when they satisfy a group identity. For torsion groups, Dooms and Ruiz Marín provided several partial results in [28]. Subsequently, in [34], Giambruno et al. generalized Theorems 2.4.8 and 2.4.9. Their main result is the following.

Theorem 7.3.8. *Let F be an infinite field of characteristic $p \neq 2$ and let G be a torsion group having an involution $*$. Let FG have the involution induced from $*$. Then we have the following.*

1. If FG is semiprime, then $\mathscr{U}^+(FG)$ satisfies a group identity if and only if G is abelian or an SLC-group.
2. If FG is not semiprime, then $\mathscr{U}^+(FG)$ satisfies a group identity if and only if the p-elements of G form a (normal) subgroup P, G has a p-abelian subgroup of finite index, and either

   a. G' is a p-group of bounded exponent, or
   b. G/P is an SLC-group and G has a normal *-invariant p-subgroup N of bounded exponent such that $P/N \leq \zeta(G/N)$, and the induced involution acts as the identity on P/N.

In particular, we note that if G is torsion, and $\mathscr{U}^+(FG)$ satisfies a group identity, then FG satisfies a polynomial identity. The classification problem remains open if G has elements of infinite order.

Making use of the preceding theorem, Lee et al. [72] generalized Theorems 4.5.6 and 4.5.7 provided F is infinite. Once again, the characteristic zero case follows immediately. If $p > 0$, then we can assume that G/P is abelian or an SLC-group. In the former case, we find that if $\mathscr{U}^+(FG)$ is nilpotent, then $\mathscr{U}(FG)$ is nilpotent. The exceptional case turns out to be

Theorem 7.3.9. *Let F be an infinite field of characteristic $p > 2$, and let G be a torsion group having an involution $*$. Let FG have the induced involution. Suppose that $\mathscr{U}(FG)$ is not nilpotent. Then $\mathscr{U}^+(FG)$ is nilpotent if and only if G is nilpotent and G has a finite normal *-invariant p-subgroup N such that G/N is an SLC-group.*

Proof. See [72, Theorem]. □

It seems, however, that it will be necessary to establish this result for finite fields if we are to extend the classifications of Chapter 5 in an obvious way. But comparing the preceding theorem with Theorems 7.3.4 and 7.3.5, and noting that for torsion groups G, $\mathscr{U}(FG)$ is nilpotent if and only if FG is Lie nilpotent (see Corollary 4.2.7), we obtain

Corollary 7.3.10. *Let G be a torsion group having an involution $*$, and let F be an infinite field, with FG having the induced involution. Then $\mathscr{U}^+(FG)$ is nilpotent if and only if $(FG)^+$ is Lie nilpotent.*

We conclude by mentioning that it is possible to take this a step further. Novikov introduced the notion of an oriented involution on FG. Let $\sigma : G \to \{\pm 1\}$ be a homomorphism. If $*$ is an involution on G, and $(\ker(\sigma))^* = \ker(\sigma)$, then we obtain an involution $*$ on FG via $(\sum_{g \in G} \alpha_g g)^* = \sum_{g \in G} \alpha_g \sigma(g) g^*$. Using this type of involution, the conditions under which $(FG)^+$ is commutative were determined in Broche Cristo and Polcino Milies [22]. The conditions under which $(FG)^-$ is commutative were explored in Broche Cristo et al. [21].

Appendix A
Some Results on Prime and Semiprime Rings

A.1 Definitions and Classical Results

The purpose of this appendix is to present the proof of Proposition 2.2.2. We will introduce the concepts needed for the proof in this section and provide the proof of the result in the next section.

Let R be a prime ring. Then let M be the set of all ordered pairs, (I, θ), where I is a nonzero (two-sided) ideal of R and $\theta : I \to R$ is a left R-module homomorphism. We define a relation $\sim$ on M as follows; namely, $(I_1, \theta_1) \sim (I_2, \theta_2)$ if there is a nonzero ideal of R contained in $I_1 \cap I_2$ upon which θ_1 and θ_2 agree. We observe that $\sim$ is an equivalence relation. Indeed, it is obvious that it is both reflexive and symmetric. If θ_1 and θ_2 agree on $0 \neq J_1 \subseteq I_1 \cap I_2$, and θ_2 and θ_3 agree on $0 \neq J_2 \subseteq I_2 \cap I_3$, then θ_1 and θ_3 agree on $J_1 \cap J_2$. Furthermore, $J_1 \cap J_2 \subseteq I_1 \cap I_3$, and since R is prime, $0 \neq J_1 J_2 \subseteq J_1 \cap J_2$. Thus, $\sim$ is transitive as well. Let $[I, \theta]$ denote the equivalence class of (I, θ), and write S_0 for the set of all of these equivalence classes.

We define operations on S_0 as follows. First,

$$[I_1, \theta_1] + [I_2, \theta_2] = [I_1 \cap I_2, \theta_1 + \theta_2].$$

Since R is prime, $I_1 \cap I_2 \neq 0$, so the sum lies in S_0. Second, let

$$[I_1, \theta_1][I_2, \theta_2] = [(I_1 \cap I_2)^2, \theta_2 \circ \theta_1].$$

Once again, $(I_1 \cap I_2)^2 \neq 0$. Furthermore, take any $r_1, r_2 \in I_1 \cap I_2$. Then

$$(\theta_2 \circ \theta_1)(r_1 r_2) = \theta_2(r_1 \theta_1(r_2)).$$

Now, $r_1 \in I_2$, hence $r_1 \theta_1(r_2) \in I_2$, so θ_2 is defined on this element, and $\theta_2 \circ \theta_1 : (I_1 \cap I_2)^2 \to R$ is indeed a left R-module homomorphism. We must check that these operations are well-defined. Suppose that $[I_i, \theta_i] = [I_i', \theta_i']$, $i = 1, 2$. Then θ_i agrees with θ_i' on a nonzero ideal $J_i \subseteq I_i \cap I_i'$. Thus, $\theta_1 + \theta_2$ agrees with $\theta_1' + \theta_2'$ on $J_1 \cap J_2 \neq 0$, so $[I_1 \cap I_2, \theta_1 + \theta_2] = [I_1' \cap I_2', \theta_1' + \theta_2']$. Similarly, if $s_1, s_2 \in J_1 \cap J_2$, then

$\theta_1(s_1s_2) = s_1\theta_1(s_2) = s_1\theta_1'(s_2)$, and $s_1\theta_1(s_2) \in J_2$, so $\theta_2(s_1\theta_1(s_2)) = \theta_2'(s_1\theta_1(s_2))$. Thus, $\theta_2 \circ \theta_1$ agrees with $\theta_2' \circ \theta_1'$ on $(J_1 \cap J_2)^2$, which is a nonzero ideal contained in $(I_1 \cap I_2)^2 \cap (I_1' \cap I_2')^2$. That is, $[(I_1 \cap I_2)^2, \theta_2 \circ \theta_1] = [(I_1' \cap I_2')^2, \theta_2' \circ \theta_1']$. Furthermore, we have

Lemma A.1.1. *Let R be a prime ring, and define S_0 as above. Then S_0 is a ring containing an isomorphic copy of R.*

Proof. We saw above that the addition and multiplication operations on S_0 are well-defined and that S_0 is closed under these operations. It is trivial to check that addition is associative and commutative. Furthermore, the additive identity is $[R, 0]$, where by 0 we mean the zero function on R. Also, $-[I, \theta] = [I, -\theta]$.

Take any $[I_i, \theta_i] \in S_0$, $i = 1, 2, 3$. Then

$$([I_1, \theta_1][I_2, \theta_2])[I_3, \theta_3] = [((I_1 \cap I_2)^2 \cap I_3)^2, \theta_3 \circ \theta_2 \circ \theta_1]$$

and

$$[I_1, \theta_1]([I_2, \theta_2][I_3, \theta_3]) = [(I_1 \cap (I_2 \cap I_3)^2)^2, \theta_3 \circ \theta_2 \circ \theta_1].$$

Since $((I_1 \cap I_2)^2 \cap I_3)^2 \cap (I_1 \cap (I_2 \cap I_3)^2)^2 \neq 0$, these are equal. Thus, multiplication is associative. The distributive laws follow similarly. Furthermore, it is easy to check that $[R, 1]$ is the multiplicative identity, where 1 is the identity function on R.

Finally, let us show that S_0 contains a copy of R. Define a function $\rho : R \to S_0$ via $\rho(r) = [R, \rho_r]$. Here, $\rho_r(s) = sr$ for all $s \in R$. It is plain that ρ_r is a left R-module homomorphism for all r and that ρ is a ring homomorphism. If $\rho(r) = 0$, then $[R, \rho_r] = [R, 0]$, so ρ_r is the zero function on a nonzero ideal I. That is, $Ir = 0$. In particular, then, $I(RrR) = 0$. Since R is prime, $r = 0$. Thus, ρ is injective, and S_0 contains an isomorphic copy of R.   □

We identify R with the image of R under ρ defined in the above proof and keep the notation ρ_r throughout. Also, we have the following easy observation.

Lemma A.1.2. *Let R be a prime ring. Then for any $s \in S_0$ there exists a nonzero ideal I of R such that $Is \subseteq R$. Furthermore, if $s \neq 0$ and I' is any nonzero ideal of R, then $I's \neq 0$.*

Proof. Take $s = [I, \theta] \in S_0$. If $a \in I$, then

$$[R, \rho_a][I, \theta] = [I^2, \theta \circ \rho_a].$$

If $b \in I^2$, then $\theta \circ \rho_a(b) = \theta(ba) = b\theta(a) = \rho_{\theta(a)}(b)$. Thus, $\theta \circ \rho_a$ agrees with $\rho_{\theta(a)}$ on I^2. That is, $[R, \rho_a][I, \theta] = [R, \rho_{\theta(a)}] \in R$, so $Is \subseteq R$.

Suppose that $s \neq 0$, I' is a nonzero ideal of R and $I's = 0$. Then, arguing as above,

$$0 = [R, \rho_a][I, \theta] = [I^2, \theta \circ \rho_a] = [R, \rho_{\theta(a)}]$$

for all $a \in I \cap I'$. As ρ is injective, $\theta(a) = 0$ for all $a \in I \cap I'$. Therefore, $s = [I, \theta] = 0$, giving us a contradiction and completing the proof.   □

Let C be the centre of S_0. We call C the extended centroid of R. Observe that $[I, \theta] \in C$ if and only if θ is an $(R - R)$-bimodule homomorphism on some nonzero ideal $J \subseteq I$. Indeed, if $[I, \theta] \in C$, then for any $r \in R$ we have $[I, \theta][R, \rho_r] = [R, \rho_r][I, \theta]$, hence $[I^2, \rho_r \circ \theta] = [I^2, \theta \circ \rho_r]$. That is, there exists a nonzero ideal $J \subseteq I^2$ such that $\rho_r(\theta(j)) = \theta(\rho_r(j))$ for all $j \in J$. In other words, $\theta(jr) = \theta(j)r$, and θ is a right R-module homomorphism on J. Of course, it is also a left R-module homomorphism. Conversely, let θ be an $(R - R)$-bimodule homomorphism on $0 \neq J \subseteq I$, and take any $[I', \theta'] \in S_0$. Then, for all $a, b \in J \cap I'$, we have

$$\theta(\theta'(ab)) = \theta(a\theta'(b)) = \theta(a)\theta'(b) = \theta'(\theta(a)b) = \theta'(\theta(ab)).$$

That is, $\theta \circ \theta'$ and $\theta' \circ \theta$ agree on $J \cap I'$ and, in particular, on $(J \cap I')^2$. But then

$$[I, \theta][I', \theta'] = [(I \cap I')^2, \theta' \circ \theta] = [(I \cap I')^2, \theta \circ \theta'] = [I', \theta'][I, \theta],$$

and $[I, \theta]$ is central.

Let $S = CR$. We call S the central closure of R.

Lemma A.1.3. *Let R be a prime ring. Then C is a field and S is a prime C-algebra.*

Proof. First let us show that S is prime. Suppose not, and let J_1 and J_2 be nonzero ideals of S with $J_1 J_2 = 0$. Taking $0 \neq s_i \in J_i$, we have $s_1 S s_2 = 0$. By the preceding lemma, there exist ideals I_i of R such that $0 \neq I_i s_i \subseteq R$. Thus, $(I_1 s_1)R(I_2 s_2) \subseteq I_1 s_1 S s_2 = 0$ and therefore, $(I_1 s_1 R)(I_2 s_2 R) = 0$. Since R is prime, we have a contradiction. Therefore, S is prime, and it remains to show that C is a field.

Take $0 \neq c \in C$. Choose an ideal I of R as in the previous lemma so that $0 \neq Ic \subseteq R$. Define $\theta : Ic \to I$ via $\theta(ac) = a$ for all $a \in I$. Suppose that $ac = bc$, with $a, b \in I$. Then $(a - b)c = 0$. As c is central, $S(a - b)ScS = 0$. Therefore, since S is prime, $a = b$. That is, θ is well-defined. Thus, θ is a left R-module homomorphism. Now, since c is central, Ic is a two-sided ideal of R. Write $c = [I', \theta']$. Then

$$[Ic, \theta][I', \theta'] = [(Ic \cap I')^2, \theta' \circ \theta].$$

If $a \in I$, then $(\theta' \circ \theta)(ac) = \theta'(a)$. However, identifying a with $[R, \rho_a]$ as usual, we have

$$ac = [R, \rho_a][I', \theta'] = [(I')^2, \theta' \circ \rho_a] = [R, \rho_{\theta'(a)}] = \theta'(a),$$

since

$$\rho_{\theta'(a)}(d) = d\theta'(a) = \theta'(da) = (\theta' \circ \rho_a)(d)$$

for all $d \in (I')^2$. That is, $\theta' \circ \theta$ acts as the identity on $(Ic \cap I')^2$, a nonzero ideal, and therefore,

$$[Ic, \theta]c = [Ic, \theta][I', \theta'] = [(Ic \cap I')^2, \theta' \circ \theta] = [R, 1] = 1.$$

Since c is central, this means that it has a two-sided inverse, which is surely central as well. Thus, C is a field, and we are done. $\square$

Now, if R satisfies a multilinear GPI $f(x_1, \ldots, x_n)$, then for any $r_i \in R$ and $c_i \in C$, we have $f(c_1 r_1, \ldots, c_n r_n) = c_1 \cdots c_n f(r_1, \ldots, r_n) = 0$, and similarly for sums of terms of the form $c_i r_i$. Thus, we see that f is a GPI for S as well. If f is nondegenerate for R, then it is surely nondegenerate for S. However, it is not immediately clear that this is an advantage, since we have simply passed from one prime ring to another. The following classical result due to Martindale shows us that this is, in fact, a major improvement.

Proposition A.1.4. *Let R be a prime ring and $S = CR$ its central closure. Then S satisfies a nondegenerate multilinear GPI if and only if S is primitive, S has an idempotent e such that eS is a minimal right ideal and eSe is a finite-dimensional division algebra over C.*

Proof. See [48, Theorem 1.3.2]. □

We can use Proposition A.1.4 to prove an interesting application, but we need an easy lemma.

Lemma A.1.5. *Let F be an infinite field. Then no vector space V over F can be written as a finite union of proper subspaces.*

Proof. Suppose that $V = \bigcup_{i=1}^{n} V_i$, where each V_i is a proper subspace and n is minimal. Take $v \in V \backslash V_n$ and fix $w \in V_n$. Then for each $\lambda \in F$, there must exist an i such that $v + \lambda w \in V_i$. Evidently $i \neq n$, otherwise V_n would contain $v + \lambda w - \lambda w = v$. As there are infinitely many such λ, there is a V_i that contains at least two such $v + \lambda w$, say $v + \lambda_1 w$ and $v + \lambda_2 w$. But then V_i contains their difference, namely $(\lambda_1 - \lambda_2)w$, and therefore w. That is, every element of V_n is contained in some V_i, $i < n$, and therefore $V = \bigcup_{i=1}^{n-1} V_i$. As n was minimal, we have a contradiction. □

We can now prove the following special case of Herstein and Small [49, Theorem 3].

Proposition A.1.6. *Let R be a prime F-algebra, where F is an infinite field. Suppose, for some $a_1, \ldots, a_n \in R$, that we have $a_1 r a_2 r \cdots r a_n = 0$ for all $r \in R$. Then some $a_i = 0$.*

Proof. Suppose that the a_i are all nonzero. Then R satisfies the GPI

$$a_1 x_1 a_2 x_1 \cdots x_1 a_n.$$

Using the linearization process described in Section 1.2, we obtain the multilinear GPI

$$\sum_{\sigma \in S_{n-1}} a_1 x_{\sigma(1)} a_2 x_{\sigma(2)} \cdots x_{\sigma(n-1)} a_n.$$

Suppose that this identity is degenerate. Then examining the $\sigma = (1)$ term, we see that R satisfies the GPI

$$a_1 x_1 a_2 x_2 \cdots x_{n-1} a_n.$$

That is, $a_1 R a_2 R \cdots R a_n = 0$. But as R is prime, this is impossible. Therefore, we have a nondegenerate multilinear GPI for R.

Let C be the extended centroid of R and $S = CR$ its central closure. As we observed above, S satisfies a nondegenerate multilinear GPI as well. Thus, by Proposition A.1.4, there exists a nonzero idempotent $e \in S$ such that eSe is a division ring. We see from Lemma A.1.2 that there exists an ideal I of R such that $0 \neq Ie \subseteq R$. Now, Ie is a left ideal of R, and by Proposition 1.4.7, Ie is not nilpotent. Thus, $IeIe \neq 0$ and, in particular, $eIe \neq 0$. Choose $a \in I$ so that $eae \neq 0$. Furthermore, $Ia \subseteq I$, hence Iae is also a left ideal of R. Also, $ae \neq 0$; hence $Iae \neq 0$, by Lemma A.1.2. Once again, Iae is not nilpotent, and we see that $eIae \neq 0$. Choose $b \in I$ such that $ebae \neq 0$.

Now, by Lemma A.1.3, S is prime. Thus, for any i, $1 \leq i \leq n$, $(SeS)(Sa_iS) \neq 0$. In particular, $eSa_i \neq 0$. If $eRa_i = 0$, then since C is central, $eSa_i = 0$, giving a contradiction. Thus, each $eRa_i \neq 0$. For each i, let $R_i = \{r \in R : era_i = 0\}$. Clearly R_i is an F-subspace of R, hence a proper subspace. By the preceding lemma, $R \neq \bigcup_{i=1}^{n} R_i$. Thus, let us take $r_1 \in R$ such that $er_1 a_i \neq 0$ for all i. Similarly, $er_1 a_i Sae \neq 0$, so $er_1 a_i Rae \neq 0$ for all i. By the same argument, we obtain $r_2 \in R$ such that $er_1 a_i r_2 ae \neq 0$ for all i.

As eSe is a division ring,

$$0 \neq (eae)(er_1 a_i r_2 ae) = eaer_1 a_i r_2 ae.$$

In particular,

$$aer_1 a_i r_2 ae \neq 0.$$

Let $d_i = aer_1 a_i r_2 ae$. Now, ae, r_1 and r_2 lie in R. Thus, $r_2 aeraer_1 \in R$ for all $r \in R$. In the expression $a_1 r a_2 r \cdots r a_n = 0$, replace r with $r_2 aeraer_1$. Then we obtain

$$a_1 r_2 aeraer_1 a_2 r_2 aeraer_1 \cdots r_2 aeraer_1 a_n = 0,$$

hence

$$aer_1 (a_1 r_2 aeraer_1 a_2 r_2 aer \cdots raer_1 a_n) r_2 ae = 0.$$

That is,

$$d_1 r d_2 r \cdots r d_n = 0.$$

But $d_i = a\lambda_i$, with $0 \neq \lambda_i \in eSe$, so $\lambda_i = e\lambda_i e$ for all i. Thus,

$$0 = d_1 b d_2 b \cdots b d_n = ae\lambda_1 ebae\lambda_2 e \cdots bae\lambda_n e.$$

Therefore,

$$(eae)(e\lambda_1 e)(ebae)(e\lambda_2 e)(ebae) \cdots (e\lambda_n e) = 0.$$

But eSe is a division ring, and we have a product of nonzero elements, which cannot be zero. This contradiction completes the proof. □

If R is a prime F-algebra satisfying a polynomial identity, then it is not necessary to construct the central closure. Indeed, we can simply take the centre Z of R and find its field of quotients, then calculate $Z^{-1}R$. The following is the sharpened version of Posner's theorem discovered independently by Formanek, Markov, Martindale, Procesi, Rowen, Schacher and Small.

Proposition A.1.7. *Let F be a field and R a prime F-algebra satisfying a polynomial identity. If Z is the centre of R, then $Z^{-1}R \cong M_n(D)$ for some division algebra D and some positive integer n.*

Proof. See [82, Theorem 5.4.10].   □

What can we say about involutions? One piece of good news is this:

Lemma A.1.8. *Let R be a prime ring with involution $*$. Then $*$ can be extended to an involution of the central closure S of R.*

Proof. Let $S = CR$ as usual. Take $s = c_1 r_1 + \cdots + c_k r_k \in S$, with $c_i \in C$ and $r_i \in R$. Using Lemma A.1.2, find a nonzero ideal I_i of R such that $I_i c_i \subseteq R$ for each i. Then letting $I = \bigcap_{i=1}^k I_i$, we see that I is a nonzero ideal of R such that $Is \subseteq R$ and, since each c_i is central, $sI \subseteq R$. Consider S_0 as above. Then, letting S_1 be the set of all $s \in S_0$ for which there exists a nonzero ideal I of R such that $sI \subseteq R$, we see that $S \subseteq S_1$.

Take any $s \in S_1$, and let I be an ideal as described above. Replacing I with $I \cap I^*$, we may assume that I is $*$-invariant. Define $\theta : I \to R$ via $\theta(a) = (sa^*)^*$ for all $a \in I$. Clearly θ respects addition. If $r \in R$, then

$$\theta(ra) = (s(ra)^*)^* = (sa^*r^*)^* = ((sa^*)r^*)^* = (r^*)^*(sa^*)^* = r\theta(a)$$

for all $a \in I$. Thus, θ is a left R-module homomorphism, and $[I, \theta] \in S_0$. Let $s^* = [I, \theta]$. (If I' is another nonzero ideal of R satisfying $sI' \subseteq R$ and $I's \subseteq R$, then constructing s^*, we obtain $[I', \theta]$, but $[I, \theta] = [I', \theta]$ under these conditions.)

Notice that if $s \in R$, then we may take $I = R$ and $\theta(a) = (sa^*)^* = as^*$, so $[I, \theta] = [R, \rho_{s^*}]$; that is, $*$ agrees with the involution on R. Also, if $s \in C$, then for any $r \in R$,

$$\theta(ar) = (s(ar)^*)^* = (sr^*a^*)^* = (r^*sa^*)^* = (sa^*)^*r = \theta(a)r$$

for all $a \in I$. Thus, θ is a right R-module homomorphism as well, hence $s^* \in C$.

Take any $s \in S_1$ and $b \in I$ as above. Then of course $bs \in R \subseteq S_1$. However,

$$s^*b^* = [I, \theta][R, \rho_{b^*}] = [I^2, \rho_{b^*} \circ \theta].$$

But

$$\rho_{b^*} \circ \theta(a) = \rho_{b^*}((sa^*)^*) = (sa^*)^*b^* = (bsa^*)^* = a(bs)^*$$

for all $a \in I^2$. Thus, $\rho_{(bs)^*}$ and $\rho_{b^*} \circ \theta$ agree on I^2, hence

$$(bs)^* = s^*b^*$$

for all $b \in I$. In particular, if $s \in S_1$, $b \in I$, then $s^*b = (b^*s)^* \in (Is)^* \subseteq R$, so $s^* \in S_1$. But if $b' \in I$ as well, then

$$b^*s^*b' = b^*(s^*((b')^*)^*) = b^*((b')^*s)^*,$$

and since this is a product of two terms in R, this means that

$$b^*s^*b' = ((b')^*sb)^* = (sb)^*b'.$$

That is, $(b^*s^* - (sb)^*)I = 0$. We claim that $b^*s^* - (sb)^* = 0$. Suppose not. Choose a nonzero ideal J of R such that $0 \neq J(b^*s^* - (sb)^*) \subseteq R$ (by Lemma A.1.2). Then $J(b^*s^* - (sb)^*)R$ is a nonzero ideal of R, and since R is prime, $IJ(b^*s^* - (sb)^*)R \neq 0$ and therefore $(IJ(b^*s^* - (sb)^*)R)^2 \neq 0$. In particular, $(b^*s^* - (sb)^*)I \neq 0$, and we have a contradiction. Thus,

$$(sb)^* = b^*s^*$$

for all $b \in I$.

Take any $s_1, s_2 \in S_1$. Then choose a nonzero ideal L of R such that $Ls_1, s_1L, Ls_2,$ s_2L, Ls_1s_2 and s_1s_2L are all contained in R, and L is $*$-invariant. (We can construct such an ideal for each of these elements, and then take the intersection, using the primeness of R.) Then for any $a \in L^2$, we have

$$a(s_1s_2)^* = (a^*)^*(s_1s_2)^* = (s_1s_2a^*)^*.$$

Now, $s_2L \subseteq R$, so $s_2L^2 \subseteq L$. Thus,

$$a(s_1s_2)^* = (s_2a^*)^*s_1^* = as_2^*s_1^*.$$

Thus, $L^2((s_1s_2)^* - s_2^*s_1^*) = 0$. By Lemma A.1.2, $(s_1s_2)^* = s_2^*s_1^*$ for all $s_1, s_2 \in S_1$.

Now, fix $s \in S$ and choose a nonzero $*$-invariant ideal L' of R such that $sL', L's,$ s^*L' and $L's^*$ are all contained in R. Then for any $a \in L$, we have

$$a(s^*)^* = (a^*)^*(s^*)^* = (s^*a^*)^* = ((as)^*)^* = as.$$

That is, $L'((s^*)^* - s) = 0$. By Lemma A.1.2, $(s^*)^* = s$. Certainly, it is easy to check that $(s_1 + s_2)^* = s_1^* + s_2^*$ for all $s_1, s_2 \in S_1$ as well. Therefore, $*$ is an involution on S_1 that extends the involution on R. But we have already seen that this involution maps C into C as well. Since C is central, it maps CR into CR. We are done. $\square$

We need more information about the involution on S. To that end, let us introduce some definitions. Let D be a division ring having an involution $*$, and let V be a left vector space over D. Then a bilinear form is a function $\langle \cdot, \cdot \rangle : V \times V \to D$ such that

$$\langle v_1 + v_2, v_3 \rangle = \langle v_1, v_3 \rangle + \langle v_2, v_3 \rangle,$$

$$\langle v_1, v_2 + v_3 \rangle = \langle v_1, v_2 \rangle + \langle v_1, v_3 \rangle$$

and

$$\langle \lambda_1 v_1, \lambda_2 v_2 \rangle = \lambda_1 \langle v_1, v_2 \rangle \lambda_2^*$$

for all $v_1, v_2, v_3 \in V$ and all $\lambda_1, \lambda_2 \in D$. The form is said to be nondegenerate if, for every $0 \neq v \in V$, there exist v_1 and v_2 in V such that $\langle v, v_1 \rangle \neq 0 \neq \langle v_2, v \rangle$.

Our interest lies in two particular types of bilinear forms. The form is said to be Hermitian if $\langle v_1, v_2 \rangle = \langle v_2, v_1 \rangle^*$ for all $v_1, v_2 \in V$. It is said to be alternate if $\langle v_1, v_2 \rangle = -\langle v_2, v_1 \rangle$. In either case, $\langle v_1, v_2 \rangle = 0$ if and only if $\langle v_2, v_1 \rangle = 0$. Let W be a subspace of V. We write $W^\perp$ for the orthogonal complement of W; that is, $W^\perp = \{v \in V : \langle v, W \rangle = 0\}$. Evidently $W^\perp$ is a subspace of V.

Lemma A.1.9. *Let V be a vector space over a division ring D with involution, and let $\langle \cdot, \cdot \rangle$ be a bilinear form that is either Hermitian or alternate. Assume that D has characteristic different from 2. If W is a finite-dimensional subspace of V, and the bilinear form is nondegenerate on W, then $V = W \oplus W^\perp$.*

Proof. Our proof is by induction on the dimension of W. If $W = 0$, then $V = W^\perp$, and there is nothing to do. Assume, therefore, that $W \neq 0$. Suppose the form is Hermitian. We claim that there exists $w \in W$ such that $\langle w, w \rangle \neq 0$. If not, then take $0 \neq w_1 \in W$. Since $\langle W, w_1 \rangle \neq 0$, take $w_2 \in W$ such that $\langle w_2, w_1 \rangle \neq 0$. Replacing w_2 with $\langle w_2, w_1 \rangle^{-1} w_2$, we have $\langle w_2, w_1 \rangle = 1$. Also, $\langle w_1, w_2 \rangle = 1^* = 1$. If $\langle w_2, w_2 \rangle \neq 0$, then the claim is proved. Otherwise,

$$\langle w_1 + w_2, w_1 + w_2 \rangle = \langle w_1, w_1 \rangle + \langle w_1, w_2 \rangle + \langle w_2, w_1 \rangle + \langle w_2, w_2 \rangle = 0 + 1 + 1 + 0 \neq 0.$$

The claim is proved.

Take $w \in W$ with $\langle w, w \rangle \neq 0$, and let W_1 be the subspace of V spanned by w. Notice that for any $v \in V$ we have

$$\langle v - \langle v, w \rangle \langle w, w \rangle^{-1} w, w \rangle = 0.$$

But then

$$v = (\langle v, w \rangle \langle w, w \rangle^{-1} w) + (v - \langle v, w \rangle \langle w, w \rangle^{-1} w) \in W_1 + W_1^\perp.$$

As the form is evidently nondegenerate on W_1, we have $W_1 \cap W_1^\perp = 0$, so $V = W_1 \oplus W_1^\perp$ and $W = W_1 \oplus (W \cap W_1^\perp)$. Furthermore, if $w' \in W \cap W_1^\perp$ satisfies $(w', W \cap W_1^\perp) = 0$, then surely $(w', W) = 0$, hence $w' = 0$. That is, the form is nondegenerate on $W \cap W_1^\perp$. Then by our inductive hypothesis, $W_1^\perp$ is the direct sum of $W \cap W_1^\perp$ and its orthogonal complement in $W_1^\perp$. We now have

$$V = W_1 \oplus (W_1^\perp \cap W) \oplus (((W \cap W_1^\perp)^\perp) \cap W_1^\perp) = W \oplus W^\perp,$$

as required.

Suppose, on the other hand, that the form is alternate. In this case, it is obvious that $\langle v, v \rangle = 0$ for all $v \in V$. Take $0 \neq w_1 \in W$, and, as above, choose $w_2 \in W$ so that $\langle w_1, w_2 \rangle = 1$ (and therefore $\langle w_2, w_1 \rangle = -1$). Then for any $v \in V$ we have

$$\langle v - \langle v, w_2 \rangle w_1 + \langle v, w_1 \rangle w_2, w_i \rangle = 0$$

for $i = 1, 2$. Letting W_1 be the subspace of V spanned by w_1 and w_2, we have

$$v = (\langle v, w_2 \rangle w_1 - \langle v, w_1 \rangle w_2) + (v - \langle v, w_2 \rangle w_1 + \langle v, w_1 \rangle w_2) \in W_1 + W_1^\perp.$$

If $\lambda_1 w_1 + \lambda_2 w_2 \in W_1^\perp$ (with $\lambda_i \in D$), then since

$$\langle \lambda_1 w_1 + \lambda_2 w_2, w_i \rangle = 0$$

for each i, we easily get $\lambda_1 = \lambda_2 = 0$. Thus, $V = W_1 \oplus W_1^\perp$. Now reason precisely as in the Hermitian case. □

Lemma A.1.10. *Let V be a vector space over a division ring D with involution, and let $\langle \cdot, \cdot \rangle$ be a nondegenerate bilinear form that is either Hermitian or alternate. Assume that D has characteristic different from 2. If W is a finite-dimensional subspace of V, then W is contained in a finite-dimensional subspace W_0 upon which the form is nondegenerate. Furthermore, if the form is alternate, then the dimension of W_0 is even.*

Proof. Our proof is by induction on the dimension of W. If $W = 0$, then the form is nondegenerate upon W and there is nothing to do, so assume $W \neq 0$. Suppose that there exists $w \in W$ with $\langle w, w \rangle \neq 0$. (Of course, the form is Hermitian in this case.) Then let W_1 be the subspace of V spanned by w. Clearly the form is nondegenerate on W_1. Thus, by Lemma A.1.9, $V = W_1 \oplus W_1^\perp$, and hence $W = W_1 \oplus (W_1^\perp \cap W)$. Clearly the form is nondegenerate upon $W_1^\perp$. By our inductive hypothesis, there exists a finite-dimensional subspace W_2 of $W_1^\perp$, containing $W_1^\perp \cap W$, upon which the form is nondegenerate. Let $W_0 = W_1 \oplus W_2$. Clearly the form is nondegenerate upon W_0. This case is complete.

Thus, we may assume that $\langle w, w \rangle = 0$ for all $w \in W$. Take $0 \neq w_1 \in W$ and find $v \in V$ such that $\langle w_1, v \rangle \neq 0$. Letting W_1 be the space spanned by w_1 and v, we see that $\dim W_1 = 2$. Furthermore, the form is nondegenerate upon W_1. Indeed, if $\lambda_1, \lambda_2 \in D$, and $\langle \lambda_1 w_1 + \lambda_2 v, w_1 \rangle = 0$, then $\lambda_2 = 0$, and if $\langle \lambda_1 w_1, v \rangle = 0$, then $\lambda_1 = 0$. Now, $\dim(W + W_1) = \dim W + 1$ and, by the preceding lemma, we have $V = W_1 \oplus W_1^\perp$ and $W + W_1 = W_1 \oplus (W_1^\perp \cap (W + W_1))$. Furthermore, the form is nondegenerate upon $W_1^\perp$. But now

$$\dim(W_1^\perp \cap (W + W_1)) = \dim W + 1 - 2 < \dim W.$$

By our inductive hypothesis, there exists a finite-dimensional subspace W_2 of $W_1^\perp$, containing $W_1^\perp \cap (W + W_1)$, upon which the form is nondegenerate. Let $W_0 = W_1 \oplus W_2$. Clearly, the form is nondegenerate upon W_0. Note in particular that if the form is alternate, then $\dim W_2$ is even, so $\dim W_0 = 2 + \dim W$ is even. We are done. □

Let V be a vector space over D with a nondegenerate bilinear form which is either Hermitian or alternate. We note that the linear transformations of V form a ring, $\mathrm{End}_D(V)$, acting upon V on the right. Take any $\theta \in \mathrm{End}_D(V)$. Then $\theta^* \in \mathrm{End}_D(V)$ is said to be the adjoint of θ if $\langle (v_1)\theta, v_2 \rangle = \langle v_1, (v_2)\theta^* \rangle$ for all $v_1, v_2 \in V$. We say that θ is continuous if it has an adjoint. Also, we say that θ has finite rank if $(V)\theta$ is finite-dimensional over D.

The following result is due to Kaplansky.

Proposition A.1.11. *Let R be a primitive ring with involution, and suppose that R has a minimal right ideal. Then there exists a left vector space V over a division ring D equipped with a nondegenerate Hermitian or alternate bilinear form, such that R is a ring of continuous linear transformations of V. Furthermore, R contains every continuous linear transformation of finite rank, and the involution on R agrees with the adjoint map on R. In fact, there exists an idempotent $e \in R$ such that $V = eR$ is a minimal right ideal of R, and $D = eRe$.*

Proof. See [48, Theorem 1.2.2]. □

Of course, the idempotent e in the above proposition is not necessarily the same idempotent found in Proposition A.1.4. But this will not impede us, due to the following well-known lemma.

Lemma A.1.12. *Let R be a prime ring, and let e_1 and e_2 be nonzero idempotents in R. If $e_1 Re_1$ and $e_2 Re_2$ are both division rings, then they are isomorphic.*

Proof. Since R is prime, $(Re_2 R)(Re_1 R) \neq 0$. In particular, there exists $a \in R$ such that $e_2 ae_1 \neq 0$. Also, $(Re_2 ae_1 R)(Re_2 R) \neq 0$, so $e_2 ae_1 Re_2 \neq 0$. But the latter is a right ideal in $e_2 Re_2$, which is a division ring. Thus, $e_2 ae_1 Re_2 = e_2 Re_2$. That is, there exists $b \in R$ such that $e_2 = e_2 ae_1 be_2 = (e_2 ae_1)(e_1 be_2)$. Let $e' = (e_1 be_2)(e_2 ae_1)$. Then $(e')^2 = e_1 be_2 (e_2 ae_1 be_2) e_2 ae_1 = e'$. Furthermore, $e_2 = e_2^2 = e_2 ae' be_2$, so $e' \neq 0$. Thus, e' is a nonzero idempotent in the division ring $e_1 Re_1$, and therefore $e' = e_1$.

We construct $\gamma : e_1 Re_1 \to e_2 Re_2$ via

$$\gamma(e_1 re_1) = e_2 ae_1 re_1 be_2.$$

This map clearly respects addition, and if $r_1, r_2 \in R$, then

$$\begin{aligned}
\gamma(e_1 r_1 e_1)\gamma(e_1 r_2 e_1) &= e_2 ae_1 r_1 e_1 be_2 ae_1 r_2 e_1 be_2 \\
&= e_2 ae_1 r_1 (e_1 be_2 ae_1) r_2 e_1 be_2 \\
&= e_2 a(e_1 r_1 e_1)(e_1 r_2 e_1) be_2 \\
&= \gamma((e_1 r_1 e_1)(e_1 r_2 e_1)).
\end{aligned}$$

Thus, γ is a ring homomorphism. Similarly, defining $\delta : e_2 Re_2 \to e_1 Re_1$ via

$$\delta(e_2 re_2) = e_1 be_2 re_2 ae_1,$$

we find that δ is a ring homomorphism. Furthermore,

$$\delta \circ \gamma(e_1 re_1) = e_1 be_2 ae_1 re_1 be_2 ae_1 = e_1 re_1$$

for all $r \in R$. That is, $\delta \circ \gamma$ is the identity function and similarly, so is $\gamma \circ \delta$. Thus, $e_1 Re_1$ and $e_2 Re_2$ are isomorphic. □

We also need a result due to Montgomery. Recall that an element of a ring is said to be regular if it is neither a left nor right zero-divisor. If R_1 is a subring of R_2, then we say that R_1 is a left (resp., right) order in R_2 if every regular element of R_1 is invertible in R_2 and every element of R_2 can be written in the form $b^{-1}a$ (resp., ab^{-1}), with $a, b \in R_1$ and b regular. We say that R_1 is an order in R_2 if it is both a left and right order.

In the following theorem, R is a prime ring with involution, C is its extended centroid and $S = CR$ is its central closure. By Lemma A.1.8, the involution on R extends to an involution on S. Suppose that S is primitive and has a minimal right ideal. Then we obtain D and V, together with a bilinear form, from Proposition A.1.11.

Proposition A.1.13. *Let R, C, S, D and V be as above. Then the following properties hold.*

1. *Suppose that the bilinear form is Hermitian, and let n be a positive integer no greater than $\dim_D V$. Then there exist a $*$-invariant subring R_n of R and an order E_n of D such that $M_n(E_n) \subseteq R_n \subseteq M_n(D)$. Furthermore, the involution on R_n comes from an involution of transpose type on $M_n(D)$.*
2. *Suppose that the bilinear form is alternate, and let n be an even integer no greater than $\dim_D V$. Then $D = C$, and there exist a $*$-invariant subring R_n of R and an order E_n of D such that R_n is isomorphic to $M_n(E_n)$. Furthermore, the involution on R_n comes from the symplectic involution on $M_n(D)$.*

In either case, let V_0 be a subspace of V of dimension n over D upon which the bilinear form is nondegenerate. Then we may choose R_n in such a way that $V_0 R_n \subseteq V_0$, and $V_0^{\perp} R_n = 0$.

Proof. See [48, Theorem 2.5.1]. □

A.2 Proof of Proposition 2.2.2

Before presenting the proof of the proposition, let us deal with the case where R is a matrix ring over a division ring.

Lemma A.2.1. *Let D be a division ring with char $D \neq 2$, and let $R = M_n(D)$, where n is a positive integer. Let $*$ be an involution on R. Take $A \in M_n(D)$ and let z be central in D. If there exists a positive integer k such that $(AB(zI_n - A)B^*)^k = 0$ for all $B \in R$, then $A = 0$ or zI_n.*

Proof. If $k = 1$, then the result is clear, since we are working in a division ring. Thus, we assume that $k \geq 2$. We may assume that $*$ is of one of the two types described in Proposition 2.1.4.

First, suppose that $*$ is an involution of transpose type. We claim that it is sufficient to assume that $n = 2$. Take any $1 \leq i < j \leq n$. Let $e = E_{ii} + E_{jj}$. Clearly e is a symmetric idempotent. Thus, eRe inherits the involution from R, and $eRe \cong M_2(D)$. Also, if $B \in R$, then $(A(eBe)(zI_n - A)(eB^*e))^k = 0$. Multiplying on the left by e, we get $((eAe)(eBe)(ze - eAe)(eB^*e))^k = 0$. By the $n = 2$ case, we conclude that $eAe = 0$

or ze. That is, letting $A = (a_{rs})$, we have $a_{ij} = a_{ji} = 0$, and either $a_{ii} = a_{jj} = 0$ or $a_{ii} = a_{jj} = z$. As this is true for all i and j, we have $A = 0$ or zI_n, as required.

Assume, therefore, that $n = 2$. As we are dealing with a nilpotent 2×2 matrix over a division ring, we must have $(AB(zI_2 - A)B^*)^2 = 0$ for all $B \in M_2(D)$. Let

$$U = \begin{pmatrix} u_{11} & 0 \\ 0 & u_{22} \end{pmatrix}$$

as in Proposition 2.1.4. Then we see that $E_{12}^* = u_{22}^{-1}u_{11}E_{21}$ and $E_{21}^* = u_{11}^{-1}u_{22}E_{12}$. Let $d = u_{22}^{-1}u_{11}$. Now, we have $(AE_{12}(zI_2 - A)E_{12}^*)^2 = 0$, hence

$$0 = E_{11}(AE_{12}(zI_2 - A)E_{12}^*)^2 = E_{11}(AE_{12}(zI_2 - A)dE_{21})^2 = (a_{11}(z - a_{22})d)^2 E_{11}.$$

As $d \neq 0$, we have $a_{11} = 0$ or $a_{22} = z$. Similarly, $(AE_{21}(zI_2 - A)E_{21}^*)^2 = 0$; hence

$$0 = E_{22}(AE_{21}(zI_2 - A)d^{-1}E_{12})^2 = (a_{22}(z - a_{11})d^{-1})^2 E_{22}.$$

Therefore, $a_{22} = 0$ or $a_{11} = z$.

First, suppose that $z = 0$. Then either $a_{11} = 0$ or $a_{22} = 0$. As $(AI_2(-A)I_2)^2 = 0$, A is nilpotent and, therefore, square-zero. Assuming that $a_{11} = 0$, we have

$$\begin{pmatrix} 0 & 0 \\ 0 & 0 \end{pmatrix} = \begin{pmatrix} 0 & a_{12} \\ a_{21} & a_{22} \end{pmatrix}^2 = \begin{pmatrix} a_{12}a_{21} & a_{12}a_{22} \\ a_{22}a_{21} & a_{21}a_{12} + a_{22}^2 \end{pmatrix}.$$

It follows immediately that $a_{22} = 0$ and either $a_{12} = 0$ or $a_{21} = 0$. If we assume instead that $a_{22} = 0$, then we again get that $a_{11} = a_{22} = 0$ and either $a_{12} = 0$ or $a_{21} = 0$.

Now suppose that $z \neq 0$. Then either $a_{11} = a_{22} = 0$ or $a_{11} = a_{22} = z$. Suppose that $a_{11} = a_{22} = 0$. Then $(AI_2(zI_2 - A)I_2)^2 = 0$, hence,

$$\begin{pmatrix} 0 & 0 \\ 0 & 0 \end{pmatrix} = \left(\begin{pmatrix} 0 & a_{12} \\ a_{21} & 0 \end{pmatrix} \begin{pmatrix} z & -a_{12} \\ -a_{21} & z \end{pmatrix} \right)^2 = \begin{pmatrix} -a_{12}a_{21} & a_{12}z \\ a_{21}z & -a_{21}a_{12} \end{pmatrix}^2.$$

Thus,

$$\begin{pmatrix} 0 & 0 \\ 0 & 0 \end{pmatrix} = E_{11} \begin{pmatrix} -a_{12}a_{21} & a_{12}z \\ a_{21}z & -a_{21}a_{12} \end{pmatrix}^2 E_{22} = \begin{pmatrix} 0 & -2a_{12}a_{21}a_{12}z \\ 0 & 0 \end{pmatrix}.$$

In particular, $a_{12} = 0$ or $a_{21} = 0$.

Thus, assume that $a_{11} = a_{22} = 0$, and either $a_{12} = 0$ or $a_{21} = 0$ (with no restriction upon z). Then $(A(E_{12} + E_{21})(zI_2 - A)(E_{12} + E_{21})^*)^2 = 0$, hence

$$\begin{aligned} 0 &= E_{11}(A(E_{12} + E_{21})(zI_2 - A)(E_{12} + E_{21})^*)^2 E_{11} \\ &= E_{11}(A(E_{12} + E_{21})(zI_2 - A)(dE_{21} + d^{-1}E_{12}))^2 E_{11} \\ &= (a_{12}d^{-1}a_{21}dz^2 + a_{12}^2da_{12}^2d)E_{11}. \end{aligned}$$

Thus, if $a_{21} = 0$, then $a_{12} = 0$. Similarly, since

$$E_{22}(A(E_{12} + E_{21})(zI_2 - A)(E_{12} + E_{21})^*)^2 E_{22} = 0,$$

we get that $a_{12} = 0$ implies $a_{21} = 0$. That is, A is the zero matrix.

In order to finish the case where $*$ has transpose type, it remains to assume that $a_{11} = a_{22} = z \neq 0$. But then $zI_2 - A$ has zeroes on the diagonal. Furthermore, if $B \in M_2(D)$, then

$$((zI_2 - A)B(zI_2 - (zI_2 - A))B^*)^3 = (zI_2 - A)B(AB^*(zI_2 - A)(B^*)^*)^2 AB^* = 0.$$

Thus, by the above considerations, $zI_2 - A = 0$, and hence $A = zI_2$. This case is complete.

Now, suppose that $*$ is the symplectic involution. Then n is even, say $n = 2m$, and D is a field. Let

$$A = \begin{pmatrix} A_{11} & A_{12} \\ A_{21} & A_{22} \end{pmatrix},$$

where each $A_{rs} \in M_m(D)$. Take any matrix $M \in M_m(D)$. Then since

$$\begin{pmatrix} 0 & M \\ 0 & 0 \end{pmatrix}^* = \begin{pmatrix} 0 & -M^t \\ 0 & 0 \end{pmatrix},$$

where t is the usual transpose, we have

$$\begin{pmatrix} A_{11} & A_{12} \\ A_{21} & A_{22} \end{pmatrix} \begin{pmatrix} 0 & M \\ 0 & 0 \end{pmatrix} \begin{pmatrix} zI_m - A_{11} & -A_{12} \\ -A_{21} & zI_m - A_{22} \end{pmatrix} \begin{pmatrix} 0 & M \\ 0 & 0 \end{pmatrix}^* = \begin{pmatrix} 0 & A_{11}MA_{21}M^t \\ 0 & A_{21}MA_{21}M^t \end{pmatrix}.$$

That is, $(A_{21}MA_{21}M^t)^k = 0$ for all $M \in M_m(D)$. But we have already dealt with $M_m(D)$ under the transpose involution. Using the $z = 0$ case, we see that $A_{21} = 0$. Similarly, making use of

$$\begin{pmatrix} 0 & 0 \\ M & 0 \end{pmatrix},$$

we find that $A_{12} = 0$.

Thus, we have

$$A = \begin{pmatrix} A_{11} & 0 \\ 0 & A_{22} \end{pmatrix}.$$

Then, for any $M \in M_m(D)$,

$$\begin{pmatrix} A_{11} & 0 \\ 0 & A_{22} \end{pmatrix} \begin{pmatrix} M & 0 \\ 0 & M \end{pmatrix} \begin{pmatrix} zI_m - A_{11} & 0 \\ 0 & zI_m - A_{22} \end{pmatrix} \begin{pmatrix} M & 0 \\ 0 & M \end{pmatrix}^*$$
$$= \begin{pmatrix} A_{11}M(zI_m - A_{11})M^t & 0 \\ 0 & A_{22}M(zI_m - A_{22})M^t \end{pmatrix}.$$

Thus,

$$(A_{11}M(zI_m - A_{11})M^t)^k = 0 = (A_{22}M(zI_m - A_{22})M^t)^k.$$

We once again make use of the result for the transpose involution to conclude that $A_{11} = 0$ or zI_m, and $A_{22} = 0$ or zI_m. It remains only to check that $A_{11} = A_{22}$. Suppose instead that $A_{11} = zI_m$ and $A_{22} = 0$. Then the matrix

$$\begin{pmatrix} zI_m & 0 \\ 0 & 0 \end{pmatrix} \begin{pmatrix} 0 & I_m \\ I_m & 0 \end{pmatrix} \left(\begin{pmatrix} zI_m & 0 \\ 0 & zI_m \end{pmatrix} - \begin{pmatrix} zI_m & 0 \\ 0 & 0 \end{pmatrix} \right) \begin{pmatrix} 0 & I_m \\ I_m & 0 \end{pmatrix}^*$$

is nilpotent. But this is

$$\begin{pmatrix} -z^2 I_m & 0 \\ 0 & 0 \end{pmatrix}$$

which is not nilpotent (unless $z = 0$, in which case we are done anyway). The same argument works if $A_{11} = 0$ and $A_{22} = zI_m$. We are done. $\qquad\square$

Finally, we have Proposition 2.2.2, restated here for convenience.

Proposition A.2.2. *Let F be an infinite field of characteristic different from 2. Let R be a semiprime F-algebra having an involution $*$ that fixes F elementwise. Take a central element z of R and a positive integer k. If $a \in R$ satisfies $(ab(z-a)b^*)^k = 0$ for all $b \in R$, then a is central in R.*

Proof. Suppose we can show that for all $r \in R$, we have $ar - ra \in I$ for every prime ideal I of R. Then, since R is semiprime, the intersection of the prime ideals is 0; thus, $ar - ra = 0$ for all $r \in R$, and a is central. Therefore, it suffices to show that $a + I$ is central in R/I for every prime ideal I.

Let I be a prime ideal, and suppose that $I^* \not\subseteq I$. Take any $d \in I^*$ and any $r \in R$. Then since $dr + (dr)^*$ is symmetric, we have $(a(dr + (dr)^*)(z - a)(dr + (dr)^*))^k = 0$. That is, $(a(dr + r^*d^*)(z - a)(dr + r^*d^*))^k = 0$. Working in the prime ring $\bar{R} = R/I$, we get $(\overline{ad}\bar{r}((z-a)d)\bar{r})^k = 0$ for all $\bar{r} \in \bar{R}$. By Proposition A.1.6, $\overline{ad} = 0$ or $\overline{(z-a)d} = 0$. Either way, $\bar{a}\bar{d}\overline{(z-a)}\bar{d} = 0$. Thus, this holds for all $\bar{d} \in (I + I^*)/I$. As $I^* \not\subseteq I$, fix $0 \neq \bar{d} \in (I + I^*)/I$. Then for any $\bar{r} \in \bar{R}$, $\bar{r}d\bar{r} \in (I + I^*)/I$; hence $\bar{a}\bar{r}\bar{d}\bar{r}\overline{(z-a)}\bar{r}d\bar{r} = 0$. By Proposition A.1.6, $\bar{a} = 0$ or $\overline{z-a} = 0$. That is, $a \in I$ or $a + I = z + I$. Either way, $a + I$ is central in R/I, as required.

Now, suppose that $I^* \subseteq I$. Then $\bar{R} = R/I$ has an induced involution and of course it still satisfies $(\bar{a}\bar{r}(\bar{z} - \bar{a})(\bar{r})^*)^k = 0$ for all $\bar{r} \in \bar{R}$. Thus, we replace R with $\bar{R}$ and assume that R is prime.

Now, R satisfies the $*$-GPI $(ax_1(z - a)x_1^*)^k = 0$. Applying the linearizing process described in Section 4.4, we see that R also satisfies the multilinear $*$-GPI

$$\begin{aligned} f(x_1, x_1^*, &\ldots, x_{2k}, x_{2k}^*) \\ &= \sum_{\sigma \in S_{2k}} ax_{\sigma(1)}(z - a)x_{\sigma(2)}^* ax_{\sigma(3)}(z - a)x_{\sigma(4)}^* \cdots ax_{\sigma(2k-1)}(z - a)x_{\sigma(2k)}^*. \end{aligned}$$

Let $g(x_1, \ldots, x_{4k})$ be the polynomial obtained by replacing each x_i^* with x_{2k+i}. If f is degenerate, then the monomial of g in which the variables

$$x_1, x_{2k+2}, x_3, x_{2k+4}, \ldots, x_{2k-1}, x_{4k}$$

occur, in that order, is a GPI for R as well. But this is simply

$$ax_1(z-a)x_{2k+2}ax_3(z-a)x_{2k+4}\cdots ax_{2k-1}(z-a)x_{4k}.$$

If it is a GPI for R, then $aR(z-a)R\cdots aR(z-a)R = 0$. Since R is prime, $a = 0$ or z. Either way, a is central, and we are done. Thus, we may assume that R satisfies a nondegenerate multilinear $*$-GPI; hence, by Proposition 4.4.2, a nondegenerate multilinear GPI.

Let C be the extended centroid of R and $S = CR$ the central closure. As we have seen, if R satisfies a nondegenerate multilinear GPI, then so does S. Thus, in view of Proposition A.1.4, S is primitive and has an idempotent e_1 such that e_1S is a minimal right ideal of S and e_1Se_1 is a finite-dimensional division algebra over C. Furthermore, by Lemma A.1.8, $*$ can be extended to an involution of S. By Proposition A.1.11, there exists an idempotent e_2 of S such that e_2S is a minimal right ideal of S, e_2Se_2 is a division ring, and letting $V = e_2S$ and $D = e_2Se_2$, V is a left vector space over D having a nondegenerate Hermitian or alternate bilinear form. Furthermore, S is a ring of continuous linear transformations of V (acting upon the right), and the involution on S agrees with the adjoint with respect to the bilinear form. Also, by Lemma A.1.12, $e_1Se_1 \cong e_2Se_2$, so D is finite-dimensional over C as well.

Take any $v \in V$. We claim that v and va are linearly dependent. Suppose not. Then in view of Lemma A.1.10, there exists a finite-dimensional subspace V_0 of V, containing v and va, upon which the bilinear form is nondegenerate. Let $n = \dim V$. If the bilinear form is alternate, then we may assume that n is even. By Proposition A.1.13, there exist an order E_n of D and a $*$-invariant subring R_n of R such that $M_n(E_n) \subseteq R_n \subseteq M_n(D)$. Furthermore, the involution on R_n comes from an involution of $M_n(D)$, $V_0R_n \subseteq V_0$ and $V_0^\perp R_n = 0$.

Let $\{v_1, \ldots, v_n\}$ be a basis for V_0. By Lemma A.1.9, $V = V_0 \oplus V_0^\perp$. Thus, for each i, $1 \le i \le n$, let $v_ia = v_i' + w_i$, with $v_i' \in V_0$ and $w_i \in V_0^\perp$. Choose $a' \in M_n(D)$ such that $v_ia' = v_i'$ for all i. For any $b \in R_n$ we have $v_iab = (v_i' + w_i)b = v_i'b$, $1 \le i \le n$. But also, $v_ia'b = v_i'b$. Thus, ab and $a'b$ agree on V_0. Furthermore, $z \in C$ (which is a field, by Lemma A.1.3), so z is also central in $D = e_2Se_2$. By the same argument, we conclude that $(z-a)b^*$ and $(z-a')b^*$ also agree on V_0. Therefore,

$$V_0(a'b(z-a')b^*)^k = 0.$$

As elements of $M_n(D)$ are determined by their action upon V_0, we have

$$(a'b(z-a')b^*)^k = 0$$

for all $b \in R_n$ and so, in particular, for all $b \in M_n(E_n)$.

Now, D is finite-dimensional over C. Let F_1 be the centre of D and F_2 a maximal subfield. Then of course $D \otimes_{F_1} F_2 \cong M_m(F_2)$ for some positive integer m (see [62, Theorem 15.8]). Therefore, $M_n(D) \subseteq M_n(D) \otimes_{F_1} F_2 \cong M_{mn}(F_2)$. By Proposition 1.2.7, $M_{mn}(F_2)$ satisfies the standard polynomial identity $s_{2mn}(x_1, \ldots, x_{2mn}) = 0$. Thus, $M_n(D)$, and therefore $M_n(E_n)$, must also satisfy this identity. Of course, E_n

is contained in a division ring, so it is prime. Thus, by Proposition A.1.7, $Z^{-1}E_n$ is a matrix ring over a division ring, where Z is the centre of E_n. But of course $Z^{-1}E_n \subseteq D$, hence $Z^{-1}E_n$ is a division ring. As E_n is an order in D, it follows immediately that $D = Z^{-1}E_n$ and $M_n(D) = Z^{-1}M_n(E_n)$ (where we identify Z with the centre of $M_n(E_n)$).

Take any $r \in M_n(D)$. Then $r = c^{-1}r'$, with $0 \neq c \in Z$ and $r' \in M_n(E_n)$. Therefore,

$$
\begin{aligned}
(a'r(z - a')r^*)^k &= (a'(c^{-1}r')(z - a')(c^{-1}r')^*)^k \\
&= (c^{-1}(c^{-1})^*)^k (a'r'(z - a')(r')^*)^k \\
&= 0.
\end{aligned}
$$

Thus, by the preceding lemma, $a' = 0$ or z. But since a and a' agree on V_0, we have $va = va' = 0$ or zv, where z is central in D. Thus, v and va are linearly dependent.

Now, if $\dim_D V = 1$, then, since S lives inside $\mathrm{End}_D(V)$, which is a division ring, and $(a(z - a))^k = 0$, we have $a = 0$ or $a = z$. Either way, a is central, and we are done. Therefore, assume that V is not one-dimensional. Take any v and w in V that are linearly independent. Then there exist $\lambda_v, \lambda_w, \lambda_{v+w} \in D$ such that $va = \lambda_v v$, $wa = \lambda_w w$ and $(v + w)a = \lambda_{v+w}(v + w)$. But then

$$
\lambda_{v+w}v + \lambda_{v+w}w = (v + w)a = va + wa = \lambda_v v + \lambda_w w.
$$

Therefore,

$$
(\lambda_{v+w} - \lambda_v)v + (\lambda_{v+w} - \lambda_w)w = 0.
$$

Since v and w are linearly independent, $\lambda_v = \lambda_{v+w} = \lambda_w$. It follows easily that λ_v has the same value, λ, for all $v \in V$. But now if $a_1 \in R$, then for any $v \in V$, $va_1a = \lambda va_1 = vaa_1$. Thus, a_1a and aa_1 agree on V. Since R is a ring of linear transformations on V, $a_1a = aa_1$. Thus, a is central. The proof is complete. $\square$

References

1. Bagiński, C.: A note on the derived length of the unit group of a modular group algebra. Comm. Algebra **30**, 4905–4913 (2002)
2. Balogh, Zs., Juhász, T.: Derived lengths of symmetric and skew symmetric elements in group algebras. JP J. Algebra Number Theory Appl. **12**, 191–203 (2008)
3. Balogh, Zs., Li, Y.: On the derived length of the group of units of a group algebra. J. Algebra Appl. **6**, 991–999 (2007)
4. Bateman, J.M.: On the solvability of unit groups of group algebras. Trans. Amer. Math. Soc. **157**, 73–86 (1971)
5. Bateman, J.M., Coleman, D.B.: Group algebras with nilpotent unit groups. Proc. Amer. Math. Soc. **19**, 448–449 (1968)
6. Beidar, K.I., Martindale, W.S., III, Mikhalev, A.V.: Rings with Generalized Identities. Dekker, New York (1996)
7. Bhandari, A.K., Passi, I.B.S.: Lie-nilpotency indices of group algebras. Bull. London Math. Soc. **24**, 68–70 (1992)
8. Billig, Y., Riley, D.M., Tasić, V.: Nonmatrix varieties and nil-generated algebras whose units satisfy a group identity. J. Algebra **190**, 241–252 (1997)
9. Bovdi, A.: On group algebras with solvable unit groups. In: Bokut', L.A., Ershov, Yu.L., Kostrikin, A.I. (eds.) Proceedings of the International Conference on Algebra. Part 1, pp. 81–90. Amer. Math. Soc., Providence (1992)
10. Bovdi, A.: Group algebras with a solvable group of units. Comm. Algebra **33**, 3725–3738 (2005)
11. Bovdi, A.: Group algebras with an Engel group of units. J. Aust. Math. Soc. **80**, 173–178 (2006)
12. Bovdi, A., Khripta, I.I.: Finite dimensional group algebras having solvable unit groups. Trans. Science Conference Uzhgorod State University, 227–233 (1974)
13. Bovdi, A., Khripta, I.I.: Group algebras of periodic groups with solvable multiplicative group (Russian). Mat. Zametki **22**, 421–432 (1977); English translation in Math. Notes **22**, 725–731 (1977)
14. Bovdi, A., Khripta, I.I.: The Engel property of the multiplicative group of a group algebra (Russian). Dokl. Akad. Nauk SSSR **314**, 18–20 (1990); English translation in Soviet Math. Dokl. **42**, 243–246 (1991)
15. Bovdi, A., Khripta, I.I.: The Engel property of the multiplicative group of a group algebra (Russian). Mat. Sb. **182**, 130–144 (1991); English translation in Math. USSR-Sb. **72**, 121–134 (1992)
16. Bovdi, A., Kurdics, J.: Lie properties of the group algebra and the nilpotency class of the group of units. J. Algebra **212**, 28–64 (1999)
17. Bovdi, V.: On symmetric units in group algebras. Comm. Algebra **29**, 5411–5422 (2001)

18. Bovdi, V., Kovács, L.G., Sehgal, S.K.: Symmetric units in modular group algebras. Comm. Algebra **24**, 803–808 (1996)
19. Broche Cristo, O.: Commutativity of symmetric elements in group rings. J. Group Theory **9**, 673–683 (2006)
20. Broche Cristo, O., Jespers, E., Polcino Milies, C., Ruiz Marín, M.: Antisymmetric elements in group rings. II. J. Algebra Appl. **8**, 115–127 (2009)
21. Broche Cristo, O., Jespers, E., Ruiz Marín, M.: Antisymmetric elements in group rings with an orientation morphism. Forum Math. **21**, 427–454 (2009)
22. Broche Cristo, O., Polcino Milies, C.: Symmetric elements under oriented involutions in group rings. Comm. Algebra **34**, 3347–3356 (2006)
23. Catino, F., Siciliano, S., Spinelli, E.: A note on the nilpotency class of the unit group of a modular group algebra. Math. Proc. R. Ir. Acad. **108**, 65–68 (2008)
24. Cliff, G.H., Sehgal, S.K.: Group rings with units torsion over the center. Manuscripta Math. **33**, 145–158 (1980/81)
25. Coelho, S.P.: Group rings with units of bounded exponent over the center. Canad. J. Math. **34**, 1349–1364 (1982)
26. Coleman, D.B., Passman, D.S.: Units in modular group rings. Proc. Amer. Math. Soc. **25**, 510–512 (1970)
27. Coleman, D.B., Sandling, R.: Mod 2 group algebras with metabelian unit groups. J. Pure Appl. Algebra **131**, 25–36 (1998)
28. Dooms, A., Ruiz Marín, M.: Symmetric units satisfying a group identity. J. Algebra **308**, 742–750 (2007)
29. Du, X.K.: The centers of a radical ring. Canad. Math. Bull. **35**, 174–179 (1992)
30. Fisher, J.L., Parmenter, M.M., Sehgal, S.K.: Group rings with solvable n-Engel unit groups. Proc. Amer. Math. Soc. **59**, 195–200 (1976)
31. Giambruno, A., Jespers, E., Valenti, A.: Group identities on units of rings. Arch. Math. **63**, 291–296 (1994)
32. Giambruno, A., Polcino Milies, C.: Unitary units and skew elements in group algebras. Manuscripta Math. **111**, 195–209 (2003)
33. Giambruno, A., Polcino Milies, C., Sehgal, S.K.: Lie properties of symmetric elements in group rings. J. Algebra **321**, 890–902 (2009)
34. Giambruno, A., Polcino Milies, C., Sehgal, S.K.: Group identities on symmetric units. J. Algebra **322**, 2801–2815 (2009)
35. Giambruno, A., Polcino Milies, C., Sehgal, S.K.: Group algebras of torsion groups and Lie nilpotence. J. Group Theory **13**, 221–231 (2010)
36. Giambruno, A., Sehgal, S.K.: Lie nilpotence of group rings. Comm. Algebra **21**, 4253–4261 (1993)
37. Giambruno, A., Sehgal, S.K.: Group algebras whose Lie algebra of skew-symmetric elements is nilpotent. In: Chin, W., Osterburg, J., Quinn, D. (eds.) Groups, Rings and Algebras, pp. 113–120. Amer. Math. Soc., Providence (2006)
38. Giambruno, A., Sehgal, S.K., Valenti, A.: Group algebras whose units satisfy a group identity. Proc. Amer. Math. Soc. **125**, 629–634 (1997)
39. Giambruno, A., Sehgal, S.K., Valenti, A.: Symmetric units and group identities. Manuscripta Math. **96**, 443–461 (1998)
40. Giambruno, A., Sehgal, S.K., Valenti, A.: Group identities on units of group algebras. J. Algebra **226**, 488–504 (2000)
41. Giambruno, A., Zaicev, M.: Polynomial Identities and Asymptotic Methods. Amer. Math. Soc, Providence (2005)
42. Gonçalves, J.Z.: Free subgroups of units in group rings. Canad. Math. Bull. **27**, 309–312 (1984)
43. Gonçalves, J.Z., Mandel, A.: Semigroup identities on units of group algebras. Arch. Math. **57**, 539–545 (1991)
44. Goodaire, E.G., Jespers, E., Polcino Milies, C.: Alternative Loop Rings. North-Holland, Amsterdam (1996)

45. Gorenstein, D.: Finite Groups, 2nd edition. (Reprint of the 1980 original). AMS Chelsea, Providence (2007)

46. Gruenberg, K.W.: Two theorems on Engel groups. Proc. Cambridge Philos. Soc. **49**, 377–380 (1953)

47. Gupta, N., Levin, F.: On the Lie ideals of a ring. J. Algebra **81**, 225–231 (1983)

48. Herstein, I.N.: Rings with Involution. Univ. of Chicago Press, Chicago (1976)

49. Herstein, I.N., Small, L.W.: Some comments on prime rings. J. Algebra **60**, 223–228 (1979)

50. Isaacs, I.M.: Character Theory of Finite Groups. (Reprint of the 1976 original). AMS Chelsea, Providence (2006)

51. Jespers, E., Ruiz Marín, M.: On symmetric elements and symmetric units in group rings. Comm. Algebra **34**, 727–736 (2006)

52. Kanel-Belov, A., Rowen, L.H.: Computational Aspects of Polynomial Identities. A.K. Peters, Wellesley (2005)

53. Kaplansky, I.: "Problems in the theory of rings" revisited. Amer. Math. Monthly **77**, 445–454 (1970)

54. Karpilovsky, G.: Unit Groups of Group Rings. Longman, Harlow (1989)

55. Khripta, I.I.: On the multiplicative group of a group ring (Russian). Thesis, Uzhgorod (1971)

56. Khripta, I.I.: The nilpotence of the multiplicative group of a group ring (Russian). Mat. Zametki **11**, 191–200 (1972); English translation in Math. Notes **11**, 119–124 (1972)

57. Knus, M.-A., Merkurjev, A., Rost, M., Tignol, J.-P.: The Book of Involutions. Amer. Math. Soc., Providence (1998)

58. Kurdics, J.: Engel properties of group algebras. I. Publ. Math. Debrecen **49**, 183–192 (1996)

59. Kurdics, J.: On group algebras with metabelian unit groups. Period. Math. Hungar. **32**, 57–64 (1996)

60. Kurdics, J.: Properties of the unit group of a nonmodular group algebra. Acta Math. Acad. Paedagog. Nyházi. **14**, 37–39 (1998)

61. Kurdics, J.: Engel properties of group algebras. II. J. Pure Appl. Algebra **133**, 179–196 (1998)

62. Lam, T.Y.: A First Course in Noncommutative Rings. Springer-Verlag, New York (1991)

63. Lambek, J.: Lectures on Rings and Modules. Blaisdell, Toronto (1966)

64. Lee, G.T.: Group rings whose symmetric elements are Lie nilpotent. Proc. Amer. Math. Soc. **127**, 3153–3159 (1999)

65. Lee, G.T.: The Lie n-Engel property in group rings. Comm. Algebra **28**, 867–881 (2000)

66. Lee, G.T.: Symmetric elements in group rings and related problems. Thesis, Univ. of Alberta (2000)

67. Lee, G.T.: Nilpotent symmetric units in group rings. Comm. Algebra **31**, 581–608 (2003)

68. Lee, G.T.: Groups whose irreducible representations have degree at most 2. J. Pure Appl. Algebra **199**, 183–195 (2005)

69. Lee, G.T., Polcino Milies, C., Sehgal, S.K.: Group rings whose symmetric units are nilpotent. J. Group Theory **10**, 685–701 (2007)

70. Lee, G.T., Sehgal, S.K., Spinelli, E.: Lie properties of symmetric elements in group rings II, J. Pure Appl. Algebra **213**, 1173–1178 (2009)

71. Lee, G.T., Sehgal, S.K., Spinelli, E.: Group algebras whose symmetric and skew elements are Lie solvable. Forum Math. **21**, 661–671 (2009)

72. Lee, G.T., Sehgal, S.K., Spinelli, E.: Nilpotency of group ring units symmetric with respect to an involution. J. Pure Appl. Algebra **214**, 1592–1597 (2010)

73. Lee, G.T., Spinelli, E.: Group rings whose symmetric units are solvable. Comm. Algebra **37**, 1604–1618 (2009)

74. Lee, G.T., Spinelli, E.: Group rings whose symmetric units generate an n-Engel group. Comm. Algebra, to appear

75. Liu, C.-H.: Group algebras with units satisfying a group identity. Proc. Amer. Math. Soc. **127**, 327–336 (1999)

76. Liu, C.-H., Passman, D.S.: Group algebras with units satisfying a group identity II. Proc. Amer. Math. Soc. **127**, 337–341 (1999)

77. Mal'cev, A.I.: Nilpotent semigroups (Russian). Ivanov. Gos. Ped. Inst. Uč. Zap. Fiz.-Mat. Nauki **4**, 107–111 (1953)
78. Mann, A., Shalev, A.: The nilpotency class of the unit group of a modular group algebra. II. Israel J. Math **70**, 267–277 (1990)
79. Motose, K., Ninomiya, Y.: On the solvability of unit groups of group rings. Math. J. Okayama Univ. **15**, 209–214 (1971/72)
80. Motose, K., Tominaga, H.: Group rings with solvable unit groups. Math. J. Okayama Univ. **15**, 37–40 (1971/72)
81. Passi, I.B.S., Passman, D.S., Sehgal, S.K.: Lie solvable group rings. Canad. J. Math. **25**, 748–757 (1973)
82. Passman, D.S.: The Algebraic Structure of Group Rings. Wiley, New York (1977)
83. Passman, D.S.: Observations on group rings. Comm. Algebra **5**, 1119–1162 (1977)
84. Passman, D.S.: Group algebras whose units satisfy a group identity. II. Proc. Amer. Math. Soc. **125**, 657–662 (1997)
85. Rao, M.A., Sandling, R.: The characterisation of modular group algebras having unit groups of nilpotency class 3. Canad. Math. Bull. **38**, 112–116 (1995)
86. Riley, D.M.: Group rings with hypercentral unit groups. Canad. J. Math. **43**, 425–434 (1991)
87. Riley, D.M.: Group algebras with units satisfying an Engel identity. Rend. Circ. Mat. Palermo (2) **49**, 540–544 (2000)
88. Riley, D.M., Wilson, M.C.: Group algebras and enveloping algebras with nonmatrix and semigroup identities. Comm. Algebra **27**, 3545–3556 (1999)
89. Robinson, D.J.S.: A Course in the Theory of Groups, 2nd edition. Springer, New York (1996)
90. Rowen, L.H.: Generalized polynomial identities. J. Algebra **34**, 458–480 (1975)
91. Rowen, L.H.: Polynomial Identities in Ring Theory. Academic Press, New York (1980)
92. Rowen, L.H.: Ring Theory. Academic Press, Boston (1991)
93. Sahai, M.: On group algebras KG with $U(KG)'$ nilpotent of class at most 2. In: Jain, S.K., Parvathi, S. (eds.) Noncommutative Rings, Group Rings, Diagram Algebras and Their Applications, pp. 165–173. Amer. Math. Soc., Providence (2008)
94. Sehgal, S.K.: Topics in Group Rings. Dekker, New York (1978)
95. Sehgal, S.K.: Units in Integral Group Rings. Longman, Harlow (1993)
96. Sehgal, S.K., Valenti, A.: Group algebras with symmetric units satisfying a group identity. Manuscripta Math. **199**, 243–254 (2006)
97. Shalev, A.: On associative algebras satisfying the Engel condition. Israel J. Math. **67**, 287–290 (1989)
98. Shalev, A.: On the conjectures concerning units in p-group algebras. Rend. Circ. Mat. Palermo (2) Suppl. **23**, 279–288 (1990)
99. Shalev, A.: The nilpotency class of the unit group of a modular group algebra. I. Israel J. Math. **70**, 257–266 (1990)
100. Shalev, A.: Meta-abelian unit groups of group algebras are usually abelian. J. Pure Appl. Algebra **72**, 295–302 (1991)
101. Shalev, A.: The nilpotency class of the unit group of a modular group algebra. III. Arch. Math. **60**, 136–145 (1993)
102. Strojnowski, A.: A note on u.p. groups. Comm. Algebra **8**, 231–234 (1980)
103. Zel'manov, E.I.: On Engel Lie algebras (Russian). Sibirsk. Mat. Zh. **29**, 112–117 (1988); English translation in Siberian Math. J. **29**, 777–781 (1989)

Index